Environment and Pollution in Colonial India

India is facing a river pollution crisis today. The origins of this crisis are commonly traced back to post-Independence economic development and urbanisation. This book, in contrast, shows that some important early roots of India's river pollution problem, and in particular the pollution of the Ganges, lie with British colonial policies on wastewater disposal during the late nineteenth and early twentieth centuries.

Analysing the two cornerstones of colonial river pollution history – the introduction of sewerage systems and the introduction of biological sewage treatment technologies in cities along the Ganges – the author examines different controversies around the proposed and actual discharge of untreated or treated sewage into the Ganges, which involved officials on different administrative levels as well as the Indian public. The analysis shows that the colonial state essentially ignored the problematic aspects of sewage disposal into rivers, which were clearly evident from European experience. Guided by colonial ideology and fiscal policy, colonial officials supported the introduction of the cheapest available sewerage technologies, which were technologies causing extensive pollution. Thus, policies on sewage disposal into the Ganges and other Indian rivers took on a definite shape around the turn of the twentieth century and acquired certain enduring features that were to exert great negative influence on the future development of river pollution in India.

A well-researched study on colonial river pollution history, this book presents an innovative contribution to South Asian environmental history. It is of interest to scholars working on colonial, South Asian and environmental history, and the colonial history of public health, science and technology.

Janine Wilhelm received her doctorate from Humboldt University Berlin in 2015. Her current research focuses on the environmental history of South Asia's rivers, the history of modern Yoga and Yoga philosophy.

Routledge Studies in South Asian History

Environment and Pollution in Colonial India

Sewerage technologies along the sacred Ganges

Janine Wilhelm

LONDON AND NEW YORK

First published 2016 by Routledge

2 Park Square, Milton Park, Abingdon, Oxfordshire OX14 4RN
52 Vanderbilt Avenue, New York, NY 10017

Routledge is an imprint of the Taylor & Francis Group, an informa business

First issued in paperback 2020

British Library Cataloguing in Publication Data
A catalogue record for this book is available from the British Library

Library of Congress Cataloging in Publication Data
Names: Wilhelm, Janine, author.Title: Environment and pollution in
Colonial India : sewerage technologies along the sacred Ganges /
Janine Wilhelm.
Description: Abingdon, Oxon : Routledge, 2016. | Series: Routledge
studies in South Asian history ; 15 | Includes bibliographical references
and index.
Identifiers: LCCN 2015046131| ISBN 9781138646124 (hardback) |
ISBN 9781315627717 (ebook)
Subjects: LCSH: Water quality management–Ganges River Watershed
(India and Bangladesh) | Sewage disposal in rivers, lakes, etc.–
Environmental aspects–Ganges River Watershed (India and Bangladesh) |
Water quality management–Government policy–India. | Sewerage–India–
Government policy. | Water–Pollution–Ganges River Watershed (India
and Bangladesh)–History–19th century. | Water–Pollution–Ganges River
Watershed (India and Bangladesh)–History–20th century.
Classification: LCC TD304.G36 W55 2016 | DDC 628.30954/1–dc23
LC record available at http://lccn.loc.gov/2015046131

ISBN: 978-1-138-64612-4 (hbk)
ISBN: 978-0-367-59669-9 (pbk)

Typeset in Times New Roman
by Wearset Ltd, Boldon, Tyne and Wear

Contents

Acknowledgements

This book is a revised version of my PhD thesis submitted at the Faculty of Humanities and Social Sciences at Humboldt University, Berlin. I would like to express my sincere gratitude towards my supervisor Prof. Dr. Michael Mann, who provided invaluable guidance and unfailing support throughout the process, while granting me the greatest creative freedom imaginable. I am also greatly indebted to my second supervisor Prof. Dr. Harald Fischer-Tiné from the Swiss Federal Institute of Technology in Zurich, who contributed to this work with many insightful comments. During my two years of employment at his chair I enjoyed a most inspiring intellectual environment that proved to be highly conducive to my research.

The German Academic Exchange Service generously supported this project in 2011 and 2012 by funding two research stays at Jawaharlal Nehru University, New Delhi, as part of its exchange programme 'A New Passage to India'. Logistic support was provided by the Graduate Centre Humanities and Social Sciences at Leipzig University, for which I would like to thank them. I would also like to thank the staff of all the archives and libraries I consulted for this work, including the indescribably kind and helpful staff at the British Library's Asian and African Studies reading room, those of the National Archives of India, the Nehru Memorial Museum and Library, the Uttar Pradesh State Archives, and the Uttar Pradesh State Archives' regional archives in Banaras. I am also deeply grateful to David Arnold, Mahesh Rangarajan, Melitta Waligora, Anna Mohr, Anita Breuer and Routledge's anonymous reviewer(s) for their inspiring commentaries at various stages, and to Dorothea Schaefter and Jillian Morrison from Routledge for their unstinting support.

Finally, this book could never have materialised without my friends and family, whose presence and ongoing encouragement helped me get through the rough phases every project of this kind faces. In particular, I would like to thank my parents, who have always stood by my side, no matter what. To them I dedicate this book.

Abbreviations

APAC	British Library, Asia, Pacific and Africa Collections
ARDPH	Annual Report of the Director of Public Health
ARSC	Annual Report of the Sanitary Commissioner
BNR	Bengal Newspaper Reports
B.S.A.	Board of Scientific Advice
CGA	Central Ganga Authority
CPCB	Central Pollution Control Board
Dpt	Department
GAP	Ganga Action Plan
GPD	Ganga Project Directorate
GoBeng	Government of Bengal
GOI	Government of India
GoNWP	Government of the North-Western Provinces of Agra and Oudh
GoPun	Government of Punjab
GoUP	Government of the United Provinces of Agra and Oudh
I.A.C.	Indian Advisory Committee
IMS	Indian Medical Service
IOR	India Office Records
NAI	National Archives of India
NGRBA	National Ganga River Basin Authority
NWP	North-Western Provinces of Agra and Oudh
NWPNR	North-Western Provinces of Agra and Oudh Newspaper Reports
Prgs	Proceedings
Secy	Secretary
UP	United Provinces of Agra and Oudh
UPSA	Uttar Pradesh State Archives
UPSA(V)	Uttar Pradesh State Archives: Regional Archives, Varanasi

Introduction

India is facing an acute river pollution crisis these days. Nearly all of the country's major rivers are burdened with immense amounts of municipal sewage, industrial effluents, solid waste and other harmful substances. Pollution, moreover, is spreading fast. According to a recent report by the Indian government's Central Pollution Control Board, the number of polluted rivers rose from 121 in 2009 to 275 in 2015, while the number of polluted river stretches more than doubled during the same period, increasing from 150 to 302.[1] A primary example for the extent of this crisis is the Ganges. Banaras, Kanpur and other riparian cities discharge over 2,600 million litres of sewage into the river daily. Most of this is raw sewage, since existing sewage treatment plants handle only a fraction of the total amount and are often inadequately operated. Additionally, tanneries, oil refineries, paper mills, pharmaceutical and other industries discharge 290 million litres of often highly toxic industrial wastes. Agricultural runoff (containing pesticide and fertiliser residues), solid waste, and waste generated in connection with religious worship count among the many other sources of pollution.[2] At the same time, the river's ability to regenerate itself steadily diminishes, as water levels keep dropping due to over-extraction and the construction of countless dams.[3] Consequently, the Ganges – officially declared as India's 'national river' by Prime Minister Manmohan Singh in 2008 – has turned into one of most polluted rivers in the world.

The slow death of India's rivers is part of the nation's much more extensive environmental crisis, a crisis that became clearly palpable after Independence and has reached alarming dimensions in the wake of economic liberalisation in 1991. Similar to other emerging nations, India suffers from massive deforestation, soil contamination, polluted and dwindling ground water resources, flood disasters, recurring bouts of smog and ever increasing amounts of garbage, to name just a number of problems. In recent decades, considerable political and public awareness has come to revolve around environmental degradation in India, both on the national and international levels, and has found expression through countless political debates, conferences, press reports and publications on the subject.[4] Early government initiatives towards environmental protection commenced during the 1970s following India's participation in the United Nations' Stockholm Conference in 1972. In the same year, the government founded the National Committee on Environmental Planning and Coordination as the first national body to coordinate environmental policies and programmes, and to make environmental concerns part of economic development. In 1980, the Department of Environment, the predecessor of today's Ministry of Environment and Forests, was set up. Moreover, the 1970s and 1980s saw the passage of a great number of environmental laws, including the Wild Life (Protection) Act of 1972, the Water (Prevention and Control of Pollution) Act of 1974, the Forest (Conservation) Act of 1980, the Air (Prevention and Control of Pollution) Act of 1981, and the Environment (Protection) Act of 1986.[5] Government initiatives went hand in hand with a growing environmental awareness among the Indian public, initially stirred by the Chipko movement and other conflicts over forests and wildlife during the 1970s.[6] Nevertheless, environmental degradation continued apace since the implementation of environmental laws remained insufficient.

From the 1980s onwards, government initiatives increasingly focussed on the pollution of the Ganges. The Ganges is a river of outstanding importance, materially as much as spiritually. For one, it acts as a lifeline for Northern India. Running from the western Himalayas across the northern Indian plains towards the Bengal Delta, it is home to a rich diversity of vegetation and animal life and sustains a vast human population with water for drinking, irrigation, industries, and other purposes. The urban centres along the river banks alone today have a combined population of about 20 million people. At the same time, the Ganges is of immense religious and symbolic value to millions of Hindus, who use its water for many rituals and choose it as receptacle for their ashes after death. The many sects and belief systems within the Hindu tradition may differ widely from each other in many ways, but the superior religious and cultural significance of the Ganges is acknowledged by each.[7] The river's powerful symbolism is aptly conveyed in a passage written by the thoroughly secular Jawaharlal Nehru in his last will, ten years before his death in 1964. After expressing his wish to have part of his ashes strewn into the Ganges at Allahabad, Nehru remarked:

> The Ganga especially is the river of India, beloved of her people, round which are intertwined her racial memories, her hopes and fears, her songs of triumph, her victories and her defeats. She has been a symbol of India's age-long culture and civilization, ever-changing, ever-flowing, and yet ever the same Ganga. [...] [T]he Ganga has been to me a symbol and a memory of the past of India, running into the present, and flowing on to the great ocean of the future.[8]

At the 1981 session of the Indian Science Congress in Varanasi, inaugurated by Nehru's daughter Indira Gandhi, scientists expressed their concern at the growing pollution of the Ganges. In response, the Prime Minister directed the Central Board for the Prevention and Control of Water Pollution to investigate the sources of pollution at different sites in collaboration with the State Pollution Control Boards of Uttar Pradesh, Bihar and Bengal, as well as the Centre for

Study of Man and Environment in Kolkata. The results of this first comprehensive survey were published in 1984, indicating the gross pollution of the river
through municipal sewage, industrial effluents, corpses and carcasses, etc.[9] In
1985, government under its new Prime Minister Rajiv Gandhi drew up an ambitious and heavily funded plan to clean the river, the 'Ganga Action Plan' (GAP),
which in 1993 was extended to the Ganges's tributaries. The GAP envisaged a
reduction of pollution levels by at least 75 per cent, primarily through the construction of treatment plants for municipal sewage and industrial effluents, the
erection of electric crematoria, the creation of low-cost sanitation facilities, and
general river-front development.[10] From its early stages, the GAP was criticised
by NGOs, the media and others for a variety of reasons, such as the plan's
strongly bureaucratic and centralised approach, the corruption involved in its
implementation, and its failure to take into account people's religious perspectives on the river, which tended to obscure the problem of environmental pollution and hinder the development of public environmental awareness. Overall, the
GAP failed to change pollution levels to any meaningful extent, despite
enormous financial investments.[11]

Following the GAP's failure to counteract pollution, the Indian government
in 2009 established the National Ganga River Basin Authority (NGRBA) and
initiated the 'Mission Clean Ganga', which is supported by the World Bank with
one billion dollars. The NGRBA has committed itself to render the Ganges completely free from municipal sewage and industrial effluents by the year 2020. At
the same time, it has expressed itself ready to adopt a more holistic approach to
achieve this goal (e.g. by addressing the crucial importance of the minimum
water flow), and to more actively involve NGO's and the public at large.[12] Most
recently, the 'Mission Clean Ganga' has received a boost by India's new Prime
Minister Narendra Modi, who on taking office in May 2014 declared the cleaning of the Ganges as one of his top priorities.[13]

Along with the Indian government's long-standing endeavours, there has
been a surge of scholarly and popular writing on India's river pollution crisis.
Since the initiation of the GAP in 1985, numerous scholars have investigated the
environmental degradation of the Ganges and other rivers, and the various
reasons responsible for it. Amongst others, they have critically analysed the government's river clean-up schemes, their perceived structural weaknesses, and the
political, social and religious factors which keep undermining their success.[14]
For Western observers, an important motive for investigation clearly is their
bewilderment at how the Ganges and other rivers can simultaneously be worshipped as goddesses and subjected to the most appalling forms of environmental pollution.[15] Two of the most important scholarly studies are Kelly D.
Alley's *On the Banks of the Ganga* (2002) and Lena Zühlke's *Verehrung und
Verschmutzung des Ganges* (2013), which both offer in-depth analyses based on
a wide range of source material, while keeping a particular focus on discourses
around purity and pollution. David L. Haberman's *River of Love* (2006) presents
an equally profound account of the environmental condition of the Yamuna,
examining, among others, how the current pollution of this sacred river affects

religious practices of worship.[16] A number of popular works, such as Julian Crandall Hollick's *Ganga* (2007), Sarandha Jain's *In Search of Yamuna* (2011), and Cheryl Colopy's *Dirty, Sacred Rivers* (2012), also furnish rich insights into India's worsening river pollution crisis, combining travelogue and analysis in engaging ways.[17] Apart from these publications, countless press reports keep track of ongoing developments and help build and maintain public awareness.[18] However, with regional Hindi newspapers there is a strong tendency to emphasise the comparatively negligible amounts of pollution caused by religious worship rather than to name the major culprits, i.e. industrial and municipal waste. This uncritical approach owes itself to a variety of reasons, including the lack of scientific knowledge among journalists and the frequent linkages between newspaper owners and industrialists.[19]

Within recent scholarly and popular environmental discourse, the origins of India's river pollution crisis have been commonly traced back to the early decades of Independence. Most certainly, this period represents an environmental watershed. Under Jawaharlal Nehru, independent India embarked on an agenda of rapid economic development, with particular emphasis on industrialisation and large infrastructure projects.[20] Like many leaders of formerly colonised nations, Nehru looked towards economic growth as the means to remove poverty, and to 'develop' India along Western standards. This overwhelming preoccupation with productivity and production left little room for environmental concerns. Industrialisation, coupled with massive population growth and urbanisation, put unprecedented pressures on the environment by accelerating the overuse and pollution of natural resources. As a result, immense amounts of human, industrial and agricultural wastes got dumped into water bodies, and most of India's major streams turned into nothing short of sewers. In 1968, this situation came to light dramatically in Bihar, when a large stretch of the Ganges at Munger caught fire due to the excess discharge of oil wastes by the Barauni Oil Refinery.[21] At around the same time, investigations into the water quality of the Ganges at Banaras and Kanpur highlighted the severe pollution caused by municipal sewage and tannery effluents.[22]

However, in order to uncover the roots of independent India's river pollution crisis we have to look more deeply into the past. As this book is going to show, post-Independence developments built on a colonial legacy of river pollution that emerged from British policies on wastewater disposal during the late nineteenth and early twentieth centuries. Even though India as a whole was predominantly rural and not industrialised at the time,[23] sources of river pollution did very much exist, since several settlements along the Ganges had developed into major urban centres by the end of the nineteenth century. Banaras, an important Hindu pilgrimage site and home to a thriving textile industry for many centuries, counted around 200,000 inhabitants by 1890.[24] Allahabad, also a major Hindu pilgrimage site and the administrative seat of the North-Western Provinces, had a population of over 170,000 by 1901.[25] And Kanpur, still a small town of less than 60,000 in 1847, developed into a city of almost the same size like Banaras in less than four decades, as the British turned it into a major military station and a centre for leather and textile industries.[26] The major metropolis, however, was Calcutta, the imperial capital until 1911 and the centrepiece of imperial commerce. At the beginning of the twentieth century, the city counted roughly 850,000 inhabitants and was home to dozens of jute mills and other factories situated within a small area along the river.[27]

Growing urban populations and expanding industries put increasing pressure on their environment. Additionally, the colonial government during the second half of the nineteenth century introduced the technology which today contributes the largest pollution load to the Ganges and other Indian rivers: municipal sewerage systems. Modern sewerage systems were first built in Great Britain in the 1840s as part of the British sanitary movement's agenda to reduce sickness and mortality rates by improving urban living conditions. The provision of clean water and the removal of all sorts of filth were viewed as paramount. Thus, sanitary reformers championed large networks of water-flushed underground sewers as the most convenient and efficient method to get rid of excreta, domestic and other wastes. However, by mixing these wastes with large amounts of water, they created a new problem: sewage. Against initial proposals to utilise sewage as a liquid fertiliser in agriculture, most British municipalities ended up discharging it into adjacent streams, and within a few years most of the nation's rivers had turned into open sewers. For many decades, river pollution and sewage disposal remained highly controversial subjects in Britain, occupying administrators and scientists alike, and causing numerous municipal and national inquiries.[28]

In British India, Bombay and Calcutta were the first to build sewerage systems in the 1860s, which primarily served the needs of European quarters. In Bombay, the sewage was conveyed directly into the sea; in Calcutta it was put into the Bidyahari river, a stream discharging into the Bay of Bengal after a short distance.[29] Elsewhere, excreta and other wastes were collected and transported to the countryside for trenching, much as they had been in Europe for centuries.[30] The advance of sewerage systems beyond the two great port cities started once the North-Western Provinces under Lieutenant-Governor Sir Auckland Colvin (1887–92) embarked upon an ambitious agenda of urban sanitary reform, declaring the introduction of water supplies and sewerage systems in their major cities a priority. Significantly, all these cities – Kanpur, Allahabad, Banaras, Lucknow and Agra – were situated on the banks of the Ganges or one of its main tributaries, which according to Auckland Colvin and other officials naturally offered themselves as sewage receptacles. The North-Western Provinces' earliest sewerage projects, devised for Banaras and Kanpur, therefore envisaged the disposal of untreated city sewage straight into the Ganges.

Sir Auckland Colvin's agenda sparked the first extensive Indian debates around river pollution.[31] Colonial officials found themselves confronted with the same questions British administrators had been facing for many decades: What impact did the discharge of untreated sewage have on river water quality? Did sewage-laden rivers present a potential health hazard for the many people who drank their water? What was the nature of the disease agents introduced by sewage? Could rivers purify themselves from sewage and disease agents?

Or was it necessary to treat sewage beforehand, and if yes, by what method? A similar dispute flared up with full force a few years later in Calcutta, when European jute factories along the Hooghly installed septic tank latrines for their workforce and discharged the resulting septic tank effluents straight into the river above and within city limits.[32] The heated controversy that ensued not only involved colonial officials but also the Indian public at large.

The debates colonial officials and members of the Indian public fought among each other drew strongly on British precedents. In India, however, one additional contentious issue was added: the question whether it was appropriate to discharge excreta-containing sewage, viewed by Hindus as a ritually highly impure substance, into a river worshipped as sacred, as the immanent form of the Hindu goddess Ganga.

While river worship has been a feature of Hindu tradition since the beginnings of the Indus valley civilization, it has become most pronounced in the case of the Ganges over time.[33] According to various scriptures, the Ganges is the immanent form of the goddess Ganga, who descended to Earth from heavenly realms. The Ramayana narrates that Ganga was called to Earth by a devotee named Bhagiratha to purify the remains of his ancestor King Sagara's 60,000 sons, who had been burnt to ashes by *ṛṣi* (sage) Kapila. Responding to Bhagiratha's prayers, Ganga fell on the matted locks of Shiva on a high peak in the Himalayas. From there she followed Bhagiratha through the North Indian plains, reached Kapila's hermitage near the Bay of Bengal where the ashes lay, and purified them by her touch. Then she flowed further and merged with the ocean.[34] Another principal account of Ganga's descent is contained in the Bhagavata and Vishnu Puranas. According to the Bhagavata Purana, Vishnu incarnated as a dwarf, Trivikrama, to free the world from the rule of the demon Bali. Trivikrama requested Bali to grant him a gift marked by three paces. When Bali agreed, Trivikrama grew to gigantic size. With the first step, he covered the entire Earth, with the second, he reached the heavens, and with the third, he pierced the roof of the universe, entering the heavenly realms where Ganga flowed. From that crevice, Ganga poured down to Earth.[35] On Earth, the Vishnu Purana states, she 'washed away the dirt, in the form of the sins of the whole of the world, by her touch, and yet, herself remained pure'.[36] Both these myths stress one major theme: the purificatory power of the goddess in her immanent form as a river, which absolves those who come into contact with her from sin. For living devotees therefore, to bathe in the Ganges and to use its water during various rituals is an essential part of their religious practice. For the dying, it is most desirable to be cremated along the Ganges's banks and have their ashes strewn into it, as the goddess purifies their souls and provides them with a smooth transition into their next life, or even guides them to liberation.[37]

The identification of the river Ganges with the goddess Ganga, and her intricate relation to notions of purity and pollution, decisively shape debates on the pollution of the river by sewage and other human-generated wastes. As British anthropologist Mary Douglas has observed, notions of dirt, pollution and uncleanness are cultural constructs, and the reason why certain things are labelled as such is that they are perceived as 'matter out of place', i.e. matter standing outside of a culturally constructed order.[38] Scholars writing on the environmental pollution of the Ganges after Indian Independence highlight the major importance of cultural understandings of 'pollution' for people's perspectives on environmental pollution and the outcome of government clean-up programmes. Thus, the English term 'pollution', which carries a wholly secular meaning oriented along scientific standards of water quality, fails to reflect the cognitive and semantic differentiations made by Hindus to identify different forms of 'pollution'. Essentially, Hindus distinguish between two main forms of 'pollution': material uncleanness and dirtiness, for which they use the terms *gandagī* and *asvacchatā*, and sacred, or ritual, impurity, denoted by the terms *apavitratā* and *aśuddhatā*. In many ritual contexts, material cleanness/uncleanness and sacred purity/impurity are closely linked, and yet something that is considered as materially clean is not necessarily considered ritually clean.[39] On the background of this multi-layered understanding of 'pollution' and the Ganges's own ascribed purity and sacredness, the environmental pollution of the river opens a complex discursive field: What substances are considered polluting, and in what way? What happens when a polluting substance is put into the Ganges, both to the goddess Ganga – who is immanent in the Ganges – and the polluting substance itself? Sewage presents a specifically problematic case in this context as it contains human excreta, which Hindus generally consider as one of the most impure substances from a ritual point of view.[40]

Kelly D. Alley and Lena Zühlke, whose studies represent the most thorough analyses of post-Independence debates, both argue that religious notions of 'pollution' strongly counteract efforts aimed at environmental preservation.[41] Most Hindus translate the Ganges's spiritual power of purification to the material plane, holding that the goddess is also capable of doing away with material pollution, while her own sacred purity remains untouched by it. Another, similarly problematic, religious perspective is the increasing tendency to distinguish between the goddess and the river, from which follows that the goddess is not affected by material pollution. The case studies presented in this book unfurl a somewhat different picture for the colonial period, showing that religious perspectives around the turn of the twentieth century tended to reinforce movements to keep the Ganges free from sewage, rather than weaken them. However, there are also remarkable continuities. Government clean-up programmes until very recently have generally ignored the relevance of religious viewpoints, and this has been a major reason for their failure according to scholars and environmentalists. Religious perspectives, they contend, have to be positively harnessed and integrated into environmental programmes, since the majority of people will not accept or understand arguments that singularly build on a scientific worldview. As we will see, this official disregard for religious perspectives, characteristic of recent debates and government programmes, perpetuates a pattern that emerged in the course of colonial debates.

Despite many disagreements, colonial policies on sewage disposal into the Ganges and other Indian rivers took on a definite shape between 1890 and 1910,

and acquired certain enduring features that set the course for the future development of river pollution in India. Essentially, the problematic aspects around
sewage disposal into rivers, so evident from European experience, were largely
ignored in the colonial context. Due to a range of ideological and financial
reasons, colonial officials supported the introduction of the cheapest available
sewerage technologies, which were technologies causing extensive pollution.
Efforts to adopt less polluting technologies and to operate them efficiently failed
for the same reasons. By the early twentieth century, cities along the Ganges
found themselves with insufficient, ill-maintained sewerage systems, and river
pollution steadily worsened on the background of expanding urban populations
and industries. At the eve of Independence, the colonial legacy of river pollution
was thus clearly set out.

The following study analyses colonial river pollution policy and its major
determining factors in detail. First, it demonstrates how the Indian government's
prolonged resistance against germ theory and waterborne disease aetiologies
caused colonial officials to deny the potential dangers involved in discharging
untreated sewage into rivers.[42] Moreover, the majority of colonial officials
believed that sewage disposal into Indian rivers had to be assessed fundamentally different from sewage disposal into British rivers. This attitude built on what
I call the 'Indian paradigm', a concept that first emerged during early debates
around an Indian river pollution law and came to be well established by the turn
of the twentieth century. The 'Indian paradigm' was moored in a long-standing
colonial ideology that categorised the Indian environment as 'tropical', and thus
as inherently different from the 'temperate' environments of Europe.[43] Thus,
advocates of sewage disposal into Indian rivers held that the harmful effects of
sewage on river water quality were reduced by the specific, 'tropical' characteristics of the Indian environment, such as the rivers' large water volumes, the hot
climate, and the intensity of sunlight. Even though the 'Indian paradigm' was
controversial and disputed by many, it decisively influenced and directed colonial policies on sewage disposal into rivers for many decades.

Second, decisions about sewerage technology were strongly determined by
colonial fiscal policies, which have been referred to as 'fiscal conservatism'.[44]
Generally, the colonial state reserved the bulk of its revenues for the military,
and to cover the ever raising home charges (providing, for example, for the pensions of British colonial officials) and the 'India debt'. To the state, urban planning and development were of little interest, unless a city held strategic and/or
commercial importance, and was home to a significant number of Europeans. In
the 1870s, the Government of India and the state governments transferred the
financial responsibility for urban sanitary infrastructures to the municipalities,
without however providing them with the means to raise adequate funds. In the
North-Western Provinces and elsewhere, the main source of municipal income
was the octroi tax,[45] which rendered municipal incomes sensitive to economic
fluctuations and made it hard to budget for sanitary expenditure. The only way to
finance expensive large-scale infrastructure projects such as waterworks and sewerage systems was to introduce new taxation and to raise loans or grants-in-aid

from the central and provincial governments. However, new taxation was prone to provoke violent protests from urban populations, while the central and provincial governments' willingness to issue loans and grants-in-aid as requested by municipalities varied.[46]

Under these circumstances, municipal governments were forced to choose the cheapest sewerage technology available. Initially, this meant the construction of waterborne sewerage systems without including purification facilities such as sewage farms, and to discharge untreated sewage into rivers below city limits. From the turn of the twentieth century, biological sewage treatment promised to offer a much cheaper and at the same time more viable alternative, enabling the purification of sewage at existing outfalls within city limits and thus obviating the need to build extensive sewer networks.[47] However, the implementation of biological treatment methods turned out to be more difficult than expected. In the North-Western Provinces, experiments with various methods failed completely, and neither the Government of India nor the provincial government were ready to defray the necessary funds for sewage treatment research and the training of a requisite staff of 'sewage specialists'. Subsequently, the province more keenly promoted sewage and sullage farms, of which some were effectively built. But these operations, too, were characterised by the lack of adequate funds and sufficient numbers of trained staff, with the consequence that farms were often ill-maintained and not capable to cope with increasing amounts of wastewater.

In Calcutta, fiscal conservatism was less of a determining factor. Here, the precipitous adoption of septic tanks by factories along the Hooghly during the early 1900s added another grave source of river pollution in the form of septic tank effluents. Driven by an ideology of 'modernisation' and 'progress',[48] and under the suasion of Calcutta's European industrial lobby, the Bengal government failed to efficiently control septic tank installations and implement measures to control river pollution. In both the provinces, the 'Indian paradigm' held a persistent influence over river pollution policy, justifying the continued discharge of untreated sewage and septic tank effluents.

The influential role played by Calcutta's European mill owners points to the fact that early colonial river pollution policy did not develop in a secluded administrative world, but was actively shaped by local political, social and economic contexts. This book therefore takes a closer look at the 'ground-level' of the city, assessing the extent to which local interest groups tried and were able to codetermine policy. Based on source availability, Banaras and Calcutta have been chosen as case studies. A major point of interest here is the role of Indian, and more specifically Hindu, citizens and administrators. How did Hindus react to the proposed discharge of sewage into the Ganges, a river viewed as the immanent form of the goddess Ganga on Earth and thus holding enormous spiritual significance in Hindu religion? Did they attempt to pressurise government into adopting a different course, e.g. to construct sewage farms? As the sources suggest, Hindu attitudes towards this issue were in no way homogenous, but the colonial government's dominant ideologies and policies were indeed repeatedly attacked by Indian politicians, journalists and riparian residents. They considered the discharge of sewage into the Ganges (generally, or at certain locations) not only as a potential health hazard, but also as a tremendous religious sacrilege because of the great amounts of excreta contained in sewage. With this, they challenged government policy as such, and, more fundamentally, questioned the sole validity of scientific definitions of 'pollution' on which government policies built. In their perspective, Hindu notions of 'pollution' carried as much, if not more, importance. Overall however, British colonial officials refused to accept the validity of religious objections, and religious protest failed to exert any lasting influence.

Even though the roots of India's present struggle with river pollution can thus be traced back to colonial rule, it would be wrong to blame today's situation entirely on the colonial state. Independent India has perpetuated many of the structural problems responsible for river pollution during the colonial period, most importantly the lack of municipal administrational and financial autonomy.[49] This continuity of colonial policies is just one among many that affect India's environmental politics today and will be discussed in more detail at the end of the book.[50]

The colonial legacy of river pollution has only very recently begun to attract some attention from scholars. Meenakshi Sharma, for instance, has rightly claimed that colonial policies of modernisation and development present a major cause for the Ganges's environmental deterioration. However, she does not provide much further information to support this statement.[51] The first to treat the subject in more detail is Awadhendra Sharan, who includes some sections on early colonial river pollution debates revolving around the North-Western Provinces in his urban environmental history of Delhi.[52] Sharan concludes that with regard to wastewater disposal, the colonial government in the nineteenth and early twentieth centuries developed

a fairly global conception of water bodies and their capacities to act as sinks, captured best in the notion of assimilative capacity – the ability of natural waters to absorb, dilute, and disperse wastes – that promoted a controlled use of receiving waters as part of waste treatment and disposal structures.

At the same time, he points out that this conception was marked by colonial difference, as the potential harm ensuing from such a practice was evaluated differently in England and India.[53] While agreeing with Sharan's major argument, this book presents a much more comprehensive analysis, tracing debates and policy-making through the central, provincial and municipal layers of the colonial administration, and analysing argumentative strategies in-depth. This approach not only reveals, for instance, the more specific argumentative strategy of the 'Indian paradigm' as against the general ideology of colonial difference, it also highlights colonial fiscal conservatism as a key factor in the continuous neglect of sewage treatment. Moreover, it shows how local interest groups, such as Hindu citizens and industrialist lobbies, actively influenced official policies, and thus points towards the important agency of the 'ground-level' of the city.

A second important study on India's colonial river pollution history is Pratik Chakrabarti's latest article on Calcutta, which investigates urban discourses around water purity and pollution after the introduction of municipal water supplies in the 1860s.[54] This presents a highly significant contribution to India's urban environmental history, and is the first to bring to light Calcutta's septic tank controversy. However, by keeping an exclusive focus on Calcutta, Chakrabarti does not situate this urban discourse within the larger context of colonial river pollution policy as it emerged from the late 1880s. Moreover, he is primarily concerned with Hindu and official viewpoints on river water 'purity' and 'pollution' and related conflicts, which, as this book is going to show, were just one among many factors to shape colonial river pollution policy.

Accordingly, this book offers the first extensive historical study on the evolution of colonial river pollution policy and its impact on the future of India's rivers. It clearly shows that individual debates around sewage disposal and river pollution in cities such as Calcutta and Banaras must not be treated as isolated discourses, but as components within an overarching process of policy formation.

The near absence of research on India's colonial river pollution history is part of one of the biggest lacunae within the historiography on South Asia's environmental history: the lack of studies on urban environments and, more specifically, pollution. Other than in the US and Europe, where urban environmental history is a well-established field of research today,[55] historians writing on environmental issues in South Asia still maintain a very strong rural bias, predominantly focussing on forestry, irrigation, land use and wild life.[56] The leading American environmental historian Joel A. Tarr lists five primary themes of urban environmental history: (i) the impacts of the built environment and human activities on the natural environment; (ii) society's response to these impacts and efforts to alleviate environmental problems; (iii) the impacts of the natural environment on the city; (iv) city-hinterland relationships; and (v) investigations of gender, class and race in regard to environmental issues.[57] Most of these issues have not or have hardly been touched upon in the context of Indian colonial cities. As per urban environmental pollution, the only monograph available till today is Awadhendra Sharan's urban environmental history of Delhi. Other than that, only a counted few articles exist that deal explicitly with the subject, including Michael Anderson's study on air pollution in Calcutta and Christine Furedy's work on wastewater disposal in the same city.[58] Moreover, Robert G. Varady in his article on land use and environmental change in the Banaras region during the nineteenth century includes a short paragraph on environmental pollution (particularly river pollution through religious practices) in Banaras city.[59] More recently, Michael Mann with his study on water supply and excreta removal in Delhi has highlighted the importance of situating colonial policies on urban sanitation within a larger environmental context, rightly pointing out that 'the history of urbanization in (British) India is constantly reduced or restricted to a history of sanitation'.[60] A similar concern has been expressed by John Broich in his study on the introduction of urban water supplies in colonial Bombay.[61] Other important recent works include Pratik Chakrabarti's article on Calcutta, Amal Das' contribution on industrial pollution in the same city, and David Arnold's lucid investigation into the evolution of colonial discourses on pollution and toxicity in Calcutta and Bombay.[62] Thus, colonial urban environmental history has remained a largely untapped field, and existing studies moreover have mostly concentrated on the major Indian cities Calcutta, Bombay, and Delhi. In contrast, this book highlights the important role played by smaller cities such as Banaras and Kanpur in the formation of colonial river pollution and sewage disposal policies.[63]

The reasons for South Asian environmental history's continued neglect of urban environments are open to speculation. For one, it seems to mirror the colonial state's own priorities. The colonial administration clearly emphasised revenue-producing agriculture and forestry over urbanisation and industrialisation, which is why administrational records produced on the former are comparatively much more copious. Guided by source availability, historians have come to buttress the colonial perspective on India as a predominantly rural landscape.[64] Another important factor directing historians' attention to rural landscapes seems to be the implicit equation of 'environment' with 'nature'. Consequently, the city is excluded from the scope of environmental history, as it is categorised as manmade, as an expression of the 'cultural' in opposition to the 'natural'. This approach, however, is problematic for several reasons. As Michael Mann notes in his latest review essay (till date the only work containing a theoretical discussion on the issue in the context of South Asian environmental history) the differentiation between the 'natural' and the 'cultural' is arbitrary. There exists hardly any 'natural' environment that has not been shaped and transformed by human interaction some way or other. Even in remote areas such as deserts and mountains, peasants, pastoralists and nomads constantly leave their imprints on 'nature'. Moreover, 'nature' itself is dynamic, continually transformed by earthquakes, volcanic action and the like, and thus far from the static, pristine entity it is often taken to be. Thus, the 'natural' and the 'cultural', and the sharp differentiation between the two, are essentially ideological and/or social constructs.[65]

A look at US environmental history can serve to give us some inputs here. Similar to South Asian environmental history today, US environmental history was initially dominated by a strong rural perspective, put forward by leading scholars such as Donald Worster and Alfred Crosby. In a programmatic article, Worster maintained that environmental history was about 'the role and place of nature in human life'. Despite this broad definition, he explicitly excluded the city from the scope of the environmental historian by drawing a line between a 'natural' and a 'cultural' sphere, at the same time admitting the somewhat arbitrary nature of this distinction. Built environments, in his view, were man-made and 'wholly expressive of culture'. Environmental history instead should concentrate on natural, i.e. rural environments and human interactions with these.[66] Worster and others reiterated the 'agro-ecological' approach in the March 1990 issue of the *Journal of American History*, which contained a roundtable of articles aimed at demarcating the field.[67] A number of urban environmental historians subsequently challenged this agenda, questioning the exclusion of built environments from the so-called 'natural' sphere. The built environment, as Martin V. Melosi put it, 'is part of the physical world' and 'interacts and sometimes blends with the natural world'. Cities, he rightly maintained, act as major modifiers of the physical environment, influencing hydraulic cycles, polluting air and water resources, transforming their hinterland through water abstraction and waste disposal, etc.[68] On similar lines, Samuel P. Hays pointed out that the city as a 'focal point of increasing human congestion' was one of the 'most promising conceptual vehicles' to investigate environmental change brought about by human pressures.[69] At around the same time, the shift from a narrow 'agro-ecological' to a more inclusive approach was heralded by William Cronon's seminal work on the city-hinterland relationships of Chicago.[70] Since then, urban environmental history has turned into an established sub-field of US environmental history, and includes a wide range of themes such as industrial pollution, environmental movements, and the transformation of the city and its hinterlands by water supply and waste disposal systems.[71] South Asian environmental history, I argue, needs to broaden its present narrow 'agro-ecological' perspective similar to how its US counterpart did in the 1990s. This book, by analysing river pollution policies in British India, contributes to this agenda of a South Asian urban environmental history, and at the same time uncovers the roots of one of India's major environmental problems today.

Before we turn to the Indian subcontinent, it is essential to start with an overview of Great Britain's own problems with sewage disposal and river pollution in the nineteenth century, which generated heated official, scientific and public controversies over many decades. The main official inquiries, debates and arguments that developed around these issues in the metropole, and the policies and laws British governments finally adopted, served as major reference points to colonial officials in India. Without this background therefore, Indian debates cannot be adequately situated.

Great Britain's river pollution problem emerged in the wake of the industrial revolution as a consequence of rapid industrial expansion and urbanisation. Between 1801 and 1901, the national percentage of urban populations rose from 30 to 80 per cent. During the same period, the nation's overall population almost quadrupled, which further contributed to urban growth.[72] This dramatic demographic change exerted an enormous pressure on urban infrastructures, which were by no means sufficiently developed and improved in the process. The results were massive overcrowding and inadequate housing, the accumulation of human, animal, and industrial waste, air and water pollution, and, as a consequence of the latter, the pollution of the intermittent drinking water supply. Thus, the Victorian city provided a formidable breeding ground for infectious and contagious diseases like typhoid, cholera, diphtheria and tuberculosis.[73] The cause of cholera and most other diseases remained unknown until the late nineteenth century. Traditionally, the emergence of disease was ascribed to local environmental factors such as weather or wind conditions. By the mid-eighteenth century a variant of this model, the miasmatic theory, became largely accepted in medical circles. It held that diseases were caused by so-called 'miasma', i.e. effluvia or gases containing noxious particles, which emanated from accumulated putrefying organic matter such as human excrements and slaughterhouse wastes. The emergence of disease was therefore intricately related to local insanitary conditions, and the remedy to it was seen in the removal of filth and foul odours. Among the most influential exponents of localist, miasmatic theories stood the German agricultural chemist Justus von Liebig, who held considerable influence in Britain, and the German chemist and hygienist Max von Pettenkofer, equally a scientist of international renown. Pettenkofer specially emphasised the importance of local soil conditions in the occurrence of disease. According to his telluric theory for instance, it was only when cholera pathogens settled in faecally contaminated soils that cholera miasmas could arise.[74]

From the early 1850s several doctors and scientists expressed serious doubts about the miasmatic orthodoxy. In a paper published in 1849, British doctor John Snow claimed that cholera was transmitted by a specific disease agent, propagated through water contaminated by excrements.[75] On the occasion of the cholera epidemic in London in 1854, Snow observed the high incidence of infections among people drawing water from a street pump in Broad Street, Soho. As further investigations were to show, the pump lay close to a sewer. On Snow's request the pump handle was locked, and the epidemic came to a sudden halt.[76] In the same year, the Italian anatomist Filippo Pacini had in fact detected and reported on the cholera bacillus, but his findings went largely unnoticed.[77] During the same period another British doctor, William Budd, repeatedly pointed to the role of polluted water in the propagation of typhoid and cholera and, by 1865, argued that cholera was propagated exclusively through the excrements of infected persons.[78] But due to the dominance of the miasmatic theory, Budd and other exponents of contagionist, waterborne explanations of disease found few supporters. Only from the late 1870s did germ theory – which essentially states that specific diseases are caused by specific micro-organisms – gain definite momentum through the work of Louis Pasteur, Robert Koch and others. Following his pioneering microbial studies in the context of beer, wine, vinegar and silk production, Pasteur in 1876 identified the anthrax bacillus and successfully developed a vaccine against the disease, adding to his earlier success in developing a vaccine against chickenpox. In 1885, a vaccine against rabies followed. Meanwhile, Koch equally discovered the anthrax bacillus in 1876, the tuberculosis bacillus in 1882, and in 1884 succeeded in isolating the cholera bacillus from a water tank in Calcutta. Up to the end of the nineteenth century, different scientists identified the microbes responsible for numerous other diseases, such as plague, typhoid and pneumonia. On the background of these breakthroughs, germ theory became widely accepted in European medical circles in the course of the 1890s.[79] For reasons discussed further ahead, the influence of miasmatic theory lasted the longest in British India.

In the late 1830s, the British government ordered surveys into the connection between unhealthy living conditions and pauperism in view of the high disease and mortality rates among the urban poor. A leading figure in this venture was

Edwin Chadwick, whose *Report on the Sanitary Condition of the Labouring Population of Great Britain* (1842) marked the beginning of a vigorous public health movement, led by politicians, bureaucrats and doctors. According to Chadwick, it was the working classes' destitute living conditions that led to physical illness, excessive mortality, and an allegedly 'immoral' life-style among them. Skilfully working out the heavy costs all this caused to the nation, Chadwick urgently called for improvements in the urban environment. Vital parts of his projected sanitary reform were the general cleansing of towns from miasma-producing filth, the adoption of clean water supplies and efficient drainage systems, and the introduction of extensive underground networks of sewers. Continuously flushed with plenty of water, the sewers were supposed to carry excreta and other waste away before they had any chance to putrefy. The large amounts of wastewater created by his new system, Chadwick suggested, could be conveyed to the countryside through pipes and sold to farmers as a liquid fertiliser, thereby generating some income to the municipalities.[80] The British government broadly followed Chadwick's recommendations and made sanitary reform and the creation of a respective legal framework a priority. Early legislation culminated in the first Public Health Act and the establishment of a General Board of Health in 1848. During the decades that followed, sanitary legislation was consecutively extended.[81]

As a consequence of urban and industrial growth Britain's rivers got rapidly converted into open sewers. Riparian towns discarded immense loads of solid and liquid wastes into their rivers, including street sweepings, municipal sewage, industrial and manufacturing wastes from paper mills, dye-works, tanneries etc., slaughterhouse wastes and animal carcasses.[82] Already in 1840, the Aire was described as 'a reservoir of poison carefully kept for the purpose of breeding a pestilence in the town', and the Wear at Durham, three decades later, as 'simply a gigantic cesspool ... emitting a stench vile enough to generate a pestilence'.[83] Reporting on the condition of the working classes in England, Friedrich Engels in 1845 vividly described the Irk at Manchester as 'a narrow, coal-black, foulsmelling stream', more stagnating than flowing, 'full of debris and refuse':

> In dry weather, a long string of the most disgusting, blackish-green, slime pools are left standing on [its right] bank, from the depths of which bubbles of miasmatic gas constantly arise and give forth a stench unendurable even on the bridge 40 or 50 feet above the surface of the stream. [...] [B]esides this, the stream itself is checked every few paces by high weirs, behind which slime and refuse accumulate and rot in thick masses.[84]

One major cause for the deterioration of British rivers during the nineteenth century was the change in the traditional system of how towns disposed of human excreta and other wastes. Traditionally, excreta, or 'night-soil' in contemporary parlance, had been collected in privy vaults or cesspools, the contents of which were emptied by 'night-soil men' and sold as manure to farmers on the countryside. In an attempt to relieve cesspools from the pressure of a growing population, London in 1815 lifted the long existing ban to discharge domestic waste into the common sewers, which until then had been strictly reserved for surface drainage. Moreover, an increasing number of households adopted the water closet from the turn of the century, which added considerable amounts of wastewater into the sewers. As a result, the Thames deteriorated rapidly, and by 1828 the complete destruction of the fishermen's trade between Putney Bridge and Greenwich was noted.[85] Things came to a head during the exceptionally hot and dry summer of 1858, when the Thames emitted a stench so unbearable that members of Parliament at Westminster were driven out of the rooms facing the river, not least out of fear of disease-producing miasma. The 'Great Stink', as the Thames's condition of that summer became known, drove a hitherto complacent Parliament into immediate action, and long-standing plans for the reconstruction of London's sewerage system were quickly passed. From 1865 onwards, two parallel main sewer conduits to the north and south of the Thames intercepted those sewers that had previously drained directly into the river, and discharged the collected sewage at Beckton and Crossness.[86]

London's policy change of 1815 pioneered a problem which was, ironically, aggravated by the national public health reforms starting with the 1840s: sewage. What had formerly been disposed of in cesspools and privy vaults were mostly human faeces and urine. Sewage, however, not only contained human excreta, but also kitchen refuse, street washings, drainings and washings from markets, stables and slaughterhouses, effluents from factories and trading establishments, and so on. To make things worse, this concoction was mixed with two hundred times its amount of water. By promoting the nation-wide construction of water supply and sewerage systems and making house connections to sewers mandatory, sanitary laws such as the Public Health Act of 1848 and the London Metropolitan Sewers Act of the same year exacerbated river pollution on an unprecedented scale.[87] At the base of this policy lay the belief in the miasmatic disease theory, according to which the removal of filth from towns was paramount. Edwin Chadwick himself admitted that it was advisable to avoid river pollution, but believed that it nevertheless was 'an evil of almost inappreciable magnitude in comparison with the ill-health occasioned by the constant retention of several hundred thousand accumulations of pollution in the most densely-peopled districts'.[88]

Over the years, an increasing number of voices came to call for reform. Anthony Wohl points to three main phases in the British public's attitude towards river pollution during the second half of the nineteenth century. From Chadwick's report in 1842 to 1857, the immediate removal of all sewage from towns was given primary importance, with little or no concern for its effects on rivers. From 1858 to the 1870s, many advocated the treatment of sewage on land or by chemical methods. From about 1870 till the end of the century, parliamentarians, sanitary and environmental reformers reached the conviction that any system of sewage disposal should refrain from polluting rivers.[89] Here, we will discuss the most important debates, arguments and policies, to which colonial officials in India extensively referred during their debates around river pollution.

The first set of controversies was concerned with London's municipal water supply. During the nineteenth century, the capital's water was provided by private water companies, who drew water from the Thames and the Lea and distributed it after filtering it through sand beds. Complaints about the quality of this water arose soon after the policy change of 1815 and led to a first inquiry into the suitability of the two rivers as a source for municipal water supplies in 1828. The Royal Commission reporting on the issue concluded that the water left much to be improved and that supplies should be taken from another source. No concrete steps however were taken into this direction.[90] Repeatedly during the century, especially in 1849–52, 1865–8, in the early 1880s and in 1892–3, opposition against the water companies' monopoly and the poor quality, high cost, and intermittent supply of their water solidified into movements for water supply reform. Official inquiries were held, during which both the water companies and the reformers employed experts to testify in their favour. This way, the debates around the capital's water supply turned into a central stage where the most eminent contemporary British experts on water and sewage questions – chemists, engineers, and physicians by profession – brought forward their views on water quality and related issues. The reformers held that London could be provided with more, cheaper, and purer water from other sources, and pointed to the dangers of drinking sewage-polluted water in view of diseases like cholera and typhoid. The companies, on the other hand, did not deny that the Thames and the Lea were polluted with sewage and other wastes above their water intakes, but argued that natural and artificial purification processes reliably protected consumers from disease. The debates ended only in 1901, when the new Metropolitan Water Board took charge of the municipal water supply.[91]

The central issue – the suitability of sewage-laden rivers as a source for London's water supply – was debated along four principal questions: (i) Are rivers able to purify themselves naturally from sewage?; (ii) What is the exact nature of the disease agents present in sewage, and how are they affected by a putative process of river self-purification?; (iii) Can these disease agents be reliably detected by the available methods of water analysis?; and (iv) Are the filters employed by the water companies able to effectively neutralise disease agents?[92] The concept of river self-purification was one of the most fiercely contested issues. As we know today, streams are indeed able to naturally purify themselves from organic pollution to a certain extent. The presence of high concentrations of organic matter encourages the growth of bacteria, fungi and other decomposers, which convert the former into more basic substances such as carbon dioxide, nitrates, sulphates and phosphates. These are further converted by algae and other plants into their protoplasm. This natural ability towards self-purification gets disturbed once the amount of organic waste discharged into a stream is too high. Excess organic matter then starts to deplete the dissolved oxygen (DO) available in the river, on which decomposers rely for their work. Oxygen depletion also directly affects the ability of fish, insects, crustaceans etc. to survive in a stream.[93]

During the nineteenth century, advocates of river self-purification based their belief mainly on the visual observation that, unlike with standing waters, pollution in river water seemed to quickly disappear. There existed few scientific data showing the extent of the process and no significant investigations into the mechanisms it involved. Nevertheless, the concept by the 1860s had come to serve as a principal defence for pollution, and was utilised by the water companies' experts during all the water supply controversies from the 1850s to 1893. Thus, they claimed that rivers purified themselves rapidly from sewage by a process of oxidation supported by dilution, during which oxygen converted organic pollutants into harmless materials, and by way of the assimilation of sewage by aquatic animals and plants.[94] The concept is aptly illustrated by a statement of Henry Letheby, health officer to the city of London from 1855 to 1873, and a main defender of the water companies since 1861:

> I am quite ready to admit that the discharge of sewage into a river is a most improper thing, but considering the powerfully oxydising influence of water upon sewage, the many agencies which are at work destroying it, the power of precipitation, the using of it up by vegetables and aquatic plants, and by fish, and above all by the power of oxydation, I think that none of the sewage discharged into the Thames can at the present moment be discovered at Hampton.[95]

The water companies' opponents did not deny the existence of an oxygenation process that converted organic matter into harmless materials, but they questioned its reliability and its significance with regard to the harmful substances present in sewage-polluted water. With reference to Liebig's fermentation theory, some argued that the fermentation occurring in sewage may itself present a danger. Reformers drew another strong support from the theories of John Snow, who had argued that disease was not caused by visible organic filth, but by minute disease agents propagated through water. The prevailing uncertainties about the nature of pathogens, and the notion that these were likely to be invisible and thus not detectable, caused many to adopt what Christopher Hamlin calls an 'analytical nihilism' during the 1860s: the conviction that disease agents (to which by now they commonly referred to as 'germs') were too minute and elusive as to be detected by the available methods of chemical water analysis.[96] Together with this 'analytical nihilism' many adopted a 'purificatory nihilism'. Since it was impossible to detect disease agents, they declared, there also existed no assured method of water purification, which in turn meant that the water companies' filtering devices might not effectively guard against disease.[97]

The most ardent spokesman against the water companies between the late 1860s and late 1880s was chemist Edward Frankland. Together with his appointment as director of the Royal College of Chemistry in London in 1865, Frankland had inherited the post as analyst of London's water supply and in this function published his results in monthly reports.[98] Since the cholera epidemic of 1866, Frankland was convinced that water was the main medium for the spread of zymotic diseases, and that the disease agents it carried were of an altogether different nature than he had hitherto assumed. He now thought them to be minute germs, or spores, which did not dissolve by dilution or under the influence of oxygen, and could neither be measured by chemical water analysis nor removed from the water by filtration. Based on this he developed the concept of 'previous sewage contamination', which declared that water once contaminated with sewage must never be used as a source of drinking water supply, because even if the sewage matter itself may have been dissolved, the disease agents may not have.[99] Between 1868 and 1869 Frankland conducted the first empirical investigations into the process of oxidative river self-purification. His analyses of water from the Thames and several other rivers led him to the conclusion that the process of oxidation of organic matter in river water was a very slow and incomplete one. He found it impossible to state how long the water must flow before sewage matter became completely oxidised, and thought that in Britain no river was long enough for the process to be thoroughly effective.[100]

Frankland's position by the 1870s had become so dominant that the positions taken by other scientists, and the analytical processes they developed or discarded, were developed directly in relation to his. His activism for water supply reform brought him into direct confrontation with the scientists employed by the water companies. Frankland's main opponent during the late 1860s was Henry Letheby. From the 1870s onwards, a number of prominent British chemists, namely Charles Meymott Tidy, William Crookes, William Odling and James Wanklyn, constituted a loosely knit faction against Frankland, and their careers, in matters of water quality, came to be dominated by this opposition. Accordingly, they overall viewed river self-purification as effective, denied the existence of germs, and upheld the reliability of chemical water analysis and water filtration.[101] In 1880, Tidy was the first to challenge Frankland's research on river self-purification by empirical means. In his paper 'River Pollution' Tidy concluded that 'the oxidation of the organic matter in sewage, when mixed with unpolluted water and allowed a certain flow, proceeds with extreme rapidity'.[102] As to the disease agents in sewage, Tidy maintained that these were as much subject to oxidation as dead organic matter. Even if germs existed – which they most likely did not – their properties were such that they were effectively destroyed in running water.[103]

The growing acceptance of germ theory after 1884 did not, as one might expect, afford any clarity. As Christopher Hamlin's analysis of the London controversy 1892/93 shows, the advance of bacteriology merely transformed the idiom of debate without resolving the key issues. While Robert Koch's discovery of the cholera bacillus in 1883 and the isolation of the typhoid bacillus through Georg Gaffky in 1884 represented major breakthroughs, a lot of questions remained open to debate. For instance, whether these germs were the *causal* agents of cholera and typhoid,

whether bacterial species were stable; whether, assuming the responsible organisms had been found, there was any way of proving their presence or absence from a water; whether they went through a resistant or spore stage; whether and to what degree sewage, decaying vegetation, river water, still water, sterilised water, and distilled water were good or bad media for the survival and multiplication of the germs of cholera and typhoid; and finally, what combination of germs, predisposition, and environment was necessary to cause an outbreak of disease.[104]

Most of these questions remained disputed well into the next century. Thus, germ theory during the early 1890s was fraught with so many controversial issues that, especially in the context of partisan water politics, it could not provide the definite answers policy-makers were looking for.[105]

River pollution moved administrators into action on the national stage as well. Confronted with mounting pressure from reformers, parliamentarians, and interest groups such as the Fisheries Protection Association, the British government set up several commissions to investigate and find remedies to the river pollution crisis. The first of these, the Royal Commission on the Sewage of Towns (est. in 1857), in its final report of 1865 declared land treatment (i.e. the treatment of sewage by running it over land or using it to irrigate farm land) as the only effective means to purify sewage and to prevent the pollution of rivers. This method subsequently was strongly supported by the Local Government Board and became the primary means of treating sewage in Britain.[106] The major royal commission to deal with river pollution and sewage disposal was the Royal Commission on Rivers Pollution Prevention (1865–74). The commission was established in response to calls for a specific river pollution law. Hitherto, pollution cases could only be tried under common law, a slow and uncertain procedure which no one except the wealthy could afford to undertake. The first Rivers Pollution Commission served from 1865 to 1868 and was made up by civil engineers Robert Rawlinson and John Thornhill Harrison and chemist John Thomas Way. The commission was to find out to what extent municipal sewage and industrial waste could be kept out of rivers and other streams, without creating injury to public health or, significantly, loss to industry. For this purpose it was to study selected river basins representative of different sections of employment and population.[107]

Like the Royal Commission on the Sewage of Towns, the first Rivers Pollution Commission in its reports strongly advocated land treatment as the only effective means to prevent river pollution.[108] The propagation of land treatment by both commissions was representative for the sewage recycling movement that gained strength during the 1840s and 1850s and culminated into extreme optimism about the effectiveness of sewage irrigation and its potential economic benefits in the 1860s.[109] Enthusiasts based their optimism on the case of Edinburgh, where the sale of grass cultivated on the formerly barren Craigentinny Meadows was said to yield between £20 to £30 of annual returns per acre. However, none of the farms established till the 1870s was able to make similar profits. This was because sewage farming was fraught with many difficulties. The nutritional value in sewage was often very low due to its strong dilution, especially when it came from towns with a large number of water closets. The number of crops that could be cultivated was limited and often came out weedy due to seeds entering sewage from urban stables and other sources. Farming operations also involved considerable costs for the initial preparation of the land and the construction of distributing systems, and required expertise in the management and supervision of the irrigation process. The greatest difficulty however was posed by the excess amount of sewage available during the winter months and during wet seasons, when the requirement for irrigation was very low and the application of too much sewage led to the oversaturation of lands. Complaints about the 'sickening' smells emanating from sewage farms were frequent and local populations were often explicitly hostile against sewage farming. On the Continent, sewage farms were overall operated more successfully, with climatic conditions more suitable to extensive irrigation.[110]

Apart from land treatment, there existed a number of mechanical and chemical sewage treatment methods, using precipitation to recover the solid contents in sewage. Similar to sewage farm enthusiasts, advocates of these methods looked towards making profits from the sale of the sludge thus gained. However, their methods were widely criticised for being too expensive and for failing to produce a valuable fertiliser, while only clarifying, not purifying the remaining liquid. Land treatment thus remained the only viable method of sewage treatment until the late nineteenth century.[111] By the 1870s, moreover, hopes for economic profit had eclipsed and given way to the more pragmatic appreciation of sewage farms as instruments for waste disposal.[112] A true alternative to land treatment only emerged when the role of micro-organisms in the purification of sewage came to be understood. Based on this new knowledge, several biological sewage treatment methods were developed from the 1880s onwards, which will be discussed in more detail in Chapter 4.

The first Rivers Pollution Commission's advocacy for land treatment was coupled with a strong opposition to the concept of river self-purification. Rawlinson and Way both acknowledged the existence of a process of oxidative river self-purification. However, similar to Edward Frankland a few years later, they maintained that the nature of the disease agents present in sewage-polluted water was unknown, therefore concluding that river self-purification could not be trusted, that chemical water analysis and filtration provided no reliable safeguards, and that once-polluted water should not be used for drinking water supplies.[113] In 1868, the first Rivers Pollution Commission was dissolved due to irreconcilable internal differences.[114] The second commission involved Edward Frankland, engineer William Denison, and agriculturist John Chalmers Morton. The viewpoints of this second commission became inextricably linked to the person of Edward Frankland; in fact, the commission turned into Frankland's mouthpiece. In the five reports the commission submitted, Frankland further developed the perspectives of Rawlinson and Way, but whereas the first commission had based its views on speculation, Frankland relied on experiment and systematic observation. On the basis of over 3,000 water analyses and detailed investigations, Frankland concluded that the process of river self-purification did work but only very slowly. Moreover, he probed into the suitability of different soils for sewage treatment by irrigation, investigated processes for the recycling of industrial wastes and laid down chemical standards for water and effluent quality. He supported sewage treatment by irrigation, but held that there ultimately existed no method which could reliably remove disease agents from water. From the latter he concluded that even water mixed with *treated* sewage was unfit for drinking purposes.[115]

Neither the first nor the second Rivers Pollution Commission gave attention to the bacteriological factor involved in river pollution, and remained solely concerned with the chemical and physical conditions of river water, as well as certain biological effects of pollution (e.g. fish mortality). While Frankland conducted his research, Louis Pasteur in France demonstrated that fermentation and putrefaction were due to living micro-organisms, a finding that could have added an entirely new perspective to the commissioner's work. But Frankland, like other British scientists, remained under the suasion of Justus von Liebig for a long time, who was bitterly opposed to theories of fermentation as propagated by Pasteur. Only in 1874 (the closing year of the investigations) did the commissioners declare that epidemic diseases such as cholera and typhoid were most likely transmitted through domestic water supplies containing living germs of disease.[116]

In 1876, the Rivers Pollution Prevention Act came into force as the result of the Rivers Pollution Commission's inquiry. However, it was far from what reformers had been hoping for. Attempts at introducing legislation had been successively undermined by industrial pressure groups, so that clauses going against industrial interests were either considerably weakened or altogether omitted.[117]

The Act was divided in six parts. Part one completely prohibited the disposal of solid manufacturing waste into rivers. Part two prohibited the discharge of sewage into rivers. However, in the case of already existing sewers and those under construction at the time of the passing of the Act, the discharge of sewage was allowed, given the responsible local sanitary authority could prove that the sewage had been purified by 'the best practicable and available means'. In case of need, the Local Government Board could grant the sanitary authority time to adopt such a process.[118] The great weakness of these provisions lay in the fact that proceedings against pollution were to be primarily initiated by local sanitary authorities, such as local boards, who in fact counted among the worst polluters themselves. Moreover, the wording 'best practicable and available means' left ample room for interpretation and ignored the investigations of the Rivers Pollution Commission into different methods of sewage treatment.[119]

Part three addressed liquid manufacturing and mining wastes, prohibiting the discharge of any 'poisonous, noxious, or polluting liquid'. Here, too, manufacturers and mine-owners using existing sewers and those under construction could avoid penalties if they purified their effluents by 'the best practicable and available means'. Moreover, in a glaring concession to industrial interests, it was maintained that in districts that were seats of manufacturing the Local Government Board would not allow sanitary authorities to take up proceedings if it found that these would inflict 'material injury [...] on the interests of such industry'.[120] Once again, the recommendations of the Rivers Pollution

Commission were ignored by employing the phrase 'poisonous, noxious, or polluting liquid', which circumvented the introduction of fixed standards of water quality. These had been most vigorously opposed by the industrial lobby as impracticable and a threat to their trade.[121] Part four specified the administration of the law, while parts five and six addressed the application of the Act in Scotland and Ireland.

Thus, the first British national law to tackle river pollution was overall weak and inefficient. With sanitary authorities reluctant to enforce the Act, prosecutions remained few, and pollution both through sewage and industrial effluents continued. Nevertheless, certain steps towards tighter pollution control were made during the late nineteenth century. For one, courts generally supported lawsuits against polluting towns sued under common law, and often ordered them to adopt some method of sewage treatment. Secondly, a number of regional river boards such as the Mersey and Irwell Joint Committee were established from the early 1890s, which held the power to regulate and enforce uniform standards for entire river basins. These boards were backed by new, more stringent regional river pollution laws. And yet, Britain's river pollution problem persisted. Efforts towards sewage treatment were often outpaced by growing urban populations, the expansion of industries, and the more and more widespread adoption of the water closet. Moreover, a lot of disagreement existed as per the right method of sewage treatment, and the operation of sewage farms and other installations was frequently hampered by technological problems.[122] By the early twentieth century, even the ocean waters around England had become contaminated, and several outbreaks of typhoid were traced to the consumption of seafood. In conclusion, as Anthony Wohl has put it, '[t]he prevention of river pollution must be viewed [...] as one of the least satisfactory chapters in the history of Victorian public health'.[123]

Chapter preview

This book is divided into two main parts. Chapters 1 to 3 focus on the North-Western Provinces and their plans to construct waterborne sewerage systems in Banaras and Kanpur between *c*.1890 and 1900. The chapters look at the role of the central, provincial, and municipal governments, respectively, and each carries a distinct theme. Chapter 1 focuses on the Government of India, where debates principally revolved around the introduction of a national river pollution law akin the British Rivers Pollution Act, and ultimately led to the issuing of general guidelines by a special committee of inquiry in 1893. Chapter 2 turns to the government of the North-Western Provinces. After the Indian government's refusal to pass a national law in 1890, a heated debate ensued at the provincial level, which primarily revolved around questions of disease aetiology. This is when the 'Indian paradigm' acquired its full form, first integrated into old miasmatic disease theories and then adapted to the growing official acceptance of germ theory. Debates at the municipal level finally, taken up in Chapter 3, revolved around two major issues: the question of funds, as the Banaras and

Kanpur Municipal Boards struggled to finance their sewerage infrastructures, and the religious question whether the discharge of sewage into the Ganges was appropriate and in what way. Both these questions not only involved the municipal administration, but also the Indian, and particularly the Hindu, public.

The second part of the book, including Chapters 4 and 5, traces the developments further into the first decade of the twentieth century, a time when both the United (formerly: North-Western) Provinces and Bengal started to experiment with different methods of biological sewage treatment. Chapter 4 sheds light on how biological sewage treatment was received in the United Provinces, and what impact it had on established river pollution and sewage disposal policies. Chapter 5 then shifts the focus to Calcutta and the Bengal government's approach towards septic tank effluent disposal into the Hooghly. After an introduction into the beginnings of the controversy, the analysis concentrates on the proceedings of the Septic Tank Committee. The Septic Tank Committee was commissioned by the Bengal government in 1904 to lead a major inquiry into the appropriateness of septic tank effluent disposal into the Hooghly, and into the motives and validity of Hindu religious opposition against this practice. The Septic Tank Committee report, together with further investigations carried out in subsequent years, directed official policy and set the trend for future developments.

The ramifications of early colonial river pollution policy are followed up in the book's conclusion, which traces developments further into the 1930s. By this time, the consequences inherent in this policy had already become apparent. In the United Provinces, existing sewerage and sewage treatment systems were rapidly falling into decay, and with neither sufficient money nor trained staff available for counter measures, river pollution was steadily on the increase. In Calcutta, investigations into the water quality of the Hooghly furnished alarming results time and again, but sticking to its earlier industry-friendly policy and refusing to validate Indian opposition, the Bengal government did not take any concrete steps to remedy the situation. India's colonial legacy of river pollution was thus clearly set out.

Finally, some notes on terminology and transliteration as used in this book are expedient. With regard to certain Indian cities, there exists a variety of possible names and romanised spellings. Since this study is concerned with events taking place during British colonial rule, the names most commonly used that time are chosen, while antiquated spellings such as 'Cawnpore' and 'Benares' are avoided, unless they appear within a quotation. Bearing in mind the most recent name change for Calcutta, 'Kolkata' will be used when referring to this city after the date of change (2001).

All the cities playing a major role in this book, except Calcutta, were part of an administrative unit that corresponds roughly to today's states of Uttar Pradesh and Uttarakhand. During the time period concerned, this unit figured under different names: North-Western Provinces of Agra and Oudh between 1856 and 1902, United Provinces of Agra and Oudh between 1902 and 1935, and United Provinces between 1935 and 1947. For the sake of convenience, it will subsequently be referred to as 'North-Western Provinces' when talking about the period up to 1902, and as 'United Provinces' when talking about the years after 1902.

Names of Hindu gods, goddesses, mythological figures, scriptures and religious festivals are treated as if they were English words, and so are the names of organisations and newspapers. An important and helpful differentiation is made between 'Ganges' and 'Ganga', the first designating the river, the second the Hindu goddess. Other Sanskrit or Hindi words are italicised and rendered according to the international alphabet of Sanskrit transliteration, while the English 's' is used as a plural marker (e.g. *ghāṭ, ghāṭs*).

Last, the sources used for this book contain a variety of terms employed to identify specific types of wastewater, the most important ones being 'drainage', 'sewage' and 'sullage'. Drainage commonly refers to surface water (mostly from rains and street washings), which was collected in open or covered surface drains. Sewage and sullage both denote wastewater created by the mixing of household and other wastes with considerable quantities of water once urban water supplies were introduced. The major difference between them is the amount of excreta they contained. Sullage was produced in cities which chose to continue to have excreta collected by sweepers and trenched outside city limits, while the wastewater was conveyed away through open or closed drains. Sewage was created when all excreta were added to the wastewater, which was then disposed of through a sewerage system, i.e. an underground network of sewers. In practice, colonial officials often used these terms interchangeably. Also, sullage often contained considerable amounts of excreta finding their way into the drains some way or other. In what follows, it will usually be clear from the context what type of wastewater is being talked about; where this is not the case, it will be specified.

Notes

1 Central Pollution Control Board, '*CPCB Envis Newsletter*', No. 1 (January – April 2015), online, www.cpcbenvis.nic.in/envis_newsletter/ENVIS%20Newsletter%20Jan%20-%20Apr%202015.pdf (accessed 24 June 2015).
2 Ministry of Environment and Forests, Central Pollution Control Board, 'Ganga water quality trend', December 2009, online, cpcb.nic.in/upload/NewItem_168_CPCB-Ganga_Trend Report-Final.pdf (accessed 13 September 2014), pp. 29–31.
3 Zühlke 2013: 111–15.
4 For an overview see Rangarajan (2007a: 157–558), and the 'citizens' reports' published by the Centre for Science and Environment in Delhi from 1982 onwards.
5 Divan and Rosencranz 2002: 33–4, 58–67; Sharma 2009: 533.
6 Gadgil and Guha 2007: 388–90; Rangarajan 2007b: xvii.
7 Sampat 1996: 24–32; Eck 1996: 137–40.
8 Quot. in Ghose 1993: 342.
9 Das Gupta 1984.
10 Shukla and Vandana 1995: 228–44.
11 Alley 2002. According to the Central Pollution Control Board, a total of 20,000 crore rupees have been spent on Indian river clean-up programmes so far (Mishra 2014).

12 Zühlke 2013: 187–93.
13 Pradhan and Sriram (2014); Mishra and Aggarwal (2014); Parsai (2014).
14 See, for example: Ahmed 1990; Alley 1994; Alley 2002; Shukla and Vandana 1995; Markandya and Murty 2000; Haberman 2006; Centre for Science and Environment 2007; Sharma 2007; Tomalin 2009; Zühlke 2013.
15 See, for example: Hollick 2007: 1; Zühlke 2013: 18–19.
16 Haberman 2006.
17 Hollick 2007; Jain 2011; Colopy 2012. See also Trojanow 2005.
18 For a recent well-written report see Mallet (2015).
19 Zühlke 2013: 383–93.
20 Guha 2000: 63–8.
21 Sinha and Ghosh 2008: 125.
22 Chakraborty *et al.* 1965: 211–19; Saxena *et al.* 1966: 270; Agarwal *et al.* 1976: 201–6. See also Priyadarshini 2009.
23 Mann 2015a: 266–7, 272–6.
24 Freitag 2010a: 1.
25 Harrison 1980: 176; Joshi 2008: 242.
26 Bellwinkel-Schempp 1982: 134–8, 152.
27 Goode 1916: 361; Sethia 1996.
28 Wohl 1983: 233–56; Schneider 2011: xx–xxx.
29 Dossal 1991; Nath and Majumdar 1990: 167–72; Headrick 1988: 152–9.
30 Mann 2007: 23.
31 For a brief account on some aspects, see Sharan (2014: 56–60).
32 Chakrabarti 2015: 197–204.
33 Darian 2001: 14; Kinsley 2005: 188; Eck 1993: 212–13. See also Danino (2010) and Strang (2006: 83–102).
34 Alley 2002: 56–60.
35 Darian 2001: 30.
36 Quot. in Alley 2002: 60.
37 King 2005: 161–2, 175–8; Eck 1993: 215–16.
38 Douglas 2002: 44–50.
39 Alley 2002: 36–49, 75–105.
40 Hauser 2011: 204–5. Ritual impurity is also ascribed to other human excretions, such as menstrual blood, urine, sweat, mucus and earwax (ibid.).
41 Alley 2002; Zühlke 2013. David L. Haberman draws similar conclusions in his study on the environmental pollution of the Yamuna (Haberman 2006).
42 A vast body of work has been published on this feature of colonial public health policy in India. See, for example: Harrison 1994: 99–138; Watts 2001; Ogawa 2000; Isaacs, 1998.
43 Arnold 2006; Driver 2004.
44 Prashad 2001: 155.
45 The octroi was a local tax levied on various articles brought into a district or city for consumption (Oldenburg 1988: 153).
46 Mann 2015a: 296–7, 301–2; Harrison 1994: 172, 176–7.
47 Schneider 2011: xvi–xxx.
48 Baber 1996: 200–12; Sharan 2011: 426. See also Mann (2004: 1–26).
49 Mann 2015a: 307, 310.
50 See, for example: D'Souza 2008: 99–104.
51 Sharma 2007: 36, 42–4.
52 Sharan 2014: 56–60. An earlier version is included in his article 'From source to sink' (Sharan 2011: 446–9).
53 Sharan 2011: 447.
54 Chakrabarti 2015.
55 Tarr 2001: 38–9; Schott 2005: 1–2.

56 For two recent states of the art see Mann (2013) and Rangarajan (2009). For a state of the art with a particular focus on water, see D'Souza (2006). The dominance of the rural perspective is also reflected by the most recent compilations on South Asian environmental history; see Chakrabarti (2007) and Kumar *et al.* (2011a).

57 Tarr 2001: 38.

58 Anderson 1995; Furedy 1987.

59 Varady 2010.

60 Mann 2007 (quote p. 3).

61 Broich 2007 (esp. p. 364). Both authors also point to the vital role cities played in the environmental transformation of their rural hinterlands. For the need to situate urban sanitary infrastructures within a wider environmental context see also Melosi (2008: 1–2).

62 Chakrabarti 2015; Das 2007; Arnold 2013.

63 Curiously, sewerage systems and their impact on the urban environment have also been widely ignored by the historiography on technology. Till date, Daniel Headrick's *Tentacles of Progress* remains the only work to contain an account on sewerage systems in British India, namely Calcutta. Headrick does include some references on river pollution and sewage treatment, but ultimately, he follows the established pattern of limiting sewerage systems to the context of colonial sanitary policy. The colonial state, his major argument goes, used sewerage systems and water supplies as tools to segregate European from 'native' quarters, thereby reinforcing social prejudices and economic disparities (Headrick 1988: 145–70). Other early contributors to the field have similarly focused on 'big' technologies, such as railways, steamships and telegraphs, but have not considered water supplies and sewerage systems (see, for example: Headrick 1981; MacLeod and Kumar 1995; Arnold 2000: 92–128; Choudhury 2010; also Adas 1989). Lately, the discipline has shifted away from 'big' towards 'small' technologies, e.g. sewing machines and bicycles, so that it is doubtful whether sewerage systems will draw any attention in the near future (see, for example: Godley 2001; Arnold and DeWald 2011). While rooted in an environmental history approach, this book by tracing the advance of sewerage systems presents a valuable contribution to the historiography of technology in British India.

64 Mann 2013: 328–9.

65 Ibid., 324–5.

66 Worster 1988: 292–3.

67 Worster 1990. All but one author shared Worster's perspective.

68 Melosi 1993: 1–23 (quotes p. 4).

69 Hays 1998: 70. For further contributions to this debate see, for example: Meisner Rosen and Tarr 1994; Tarr and Stine 1994.

70 Cronon 1991.

71 For current themes see US environmental history's major journal *Environment and History*. Other than its US counterpart, European environmental history never showed an 'agro-ecological' bias. In a region long characterised by a thorough domestication of nature, the notions of 'untouched nature' or 'wilderness' had little relevance. Dominant themes of European environmental history are urban pollution (of air, water, soil), and environmental protest and regulation connected to it. Moreover, urban water cycles have attracted a lot of scholarly attention (Schott 2005: 4–5, 7, 18). For current themes see also the discipline's major journal *Environmental History*.

72 Halliday 2007: 18–19.

73 Wohl 1983.

74 Waller 2004: 54–7; Hardy 2005: 124–8.

75 Snow 1849: 8–12.

76 Vinten-Johansen 2003: 283–317.

77 Mann 2015b: 393–4. This was mainly due to the fact that Pacini worked in Habsburgian Italy, where he lacked institutional support (ibid.).
78 Halliday 2007: 68–76.
79 Waller 2004: 103–72.
80 Chadwick 1965.
81 Hardy 2001: 29–39.
82 Wohl 1983: 233–8. On the similar plight of rivers and other water bodies on the Continent and in the US see, for example: Büschenfeld 1997; Neri Serneri 2002; Oosthoek 2002; Benidickson 2007.
83 Quot. in Wohl 1983: 235–6.
84 Engels 2009: 62.
85 Halliday 2001: 28–9, 42–5; Palmer 1973: 22–5. The old London sewers were often natural streams that had been covered over (Halliday 2001: 28).
86 Halliday 2001: 71–102.
87 Wohl 1983: 238–9, 243–4; Halliday 2001: 47–9; Schneider 2011: xxi.
88 Chadwick 1965: 120.
89 Wohl 1983: 243.
90 Halliday 2001: 21–5.
91 Hamlin 1988a: 112; Hamlin 1987: 364–7. This and the following sections draw strongly on Christopher Hamlin, who has produced the most thorough body of research on sewage and water quality issues in Victorian Britain.
92 Hamlin 1987: 253–6.
93 Spellman 1996: 69–70. The amount of dissolved oxygen (DO) is inversely related to the biochemical oxygen demand (BOD). Both, DO and BOD, are common parameters taken to indicate the health of a river (ibid.).
94 Hamlin 1987: 253–6.
95 Henry Letheby, testifying before the Richmond Commission in May 1867. Quot. in Hamlin 1988a: 120.
96 Chemical water analysis commonly assessed water quality by measuring hardness and the amount of solids, chlorine, ammonia, nitrates, oxygen and heavy metals. However, there existed no agreed purity standards, and British chemists strongly disagreed over different methods of measurement (Hamlin 1990: 215–18).
97 Hamlin 1987: 256–93.
98 The appointment of a chemist as water analyst rested on the traditional association of chemistry with water analysis. Initially applied to natural waters from spas, wells and streams, the usage of chemical analysis gradually expanded in the course of the nineteenth century, first to the analysis of natural waters in industry and agriculture and then to the analysis of urban water supplies. For this reason, most of the experts involved in the debates around London's municipal water supply were chemists (Russell 1996: 219–22, 362–9).
99 Hamlin 1988a: 116–18.
100 Hamlin 1987: 370–6.
101 Hamlin 1990: ch. 7. Among these four, Charles Meymott Tidy and James Wanklyn were Frankland's main adversaries. Tidy, a physician and chemistry professor, succeeded Henry Letheby both as health officer to London and as main defender of the water companies after the latter's death in 1876. James Wanklyn, lecturer and water analyst in public and private practice, was Frankland's most bitter and persistent, yet also substantial, critic. As late as 1906, he refused to acknowledge the germ theory of disease, as it was too much part of Frankland's system. William Crookes, editor of the influential *Chemical News* and member of the Cattle Plague Commission in the 1860s, in fact had been among the early developers of a germ theory in relation to the cattle plague, but remained convinced that germs were air-borne. His opposition to Frankland was in part based on his deep involvement with the Native Guano Company, a commercial enterprise propagating a chemical method of sewage

treatment which Frankland openly attacked as useless. William Odling, finally, was one of the leading chemical theorists in Britain and professor of chemistry at Oxford. He comparatively remained in the background during the water controversies, but joined Tidy as analyst to the water companies in the early 1880s (ibid.).

102 Quot. in Hamlin 1987: 411.
103 Ibid.: 415–18.
104 Hamlin 1988a: 122.
105 Ibid.: 127.
106 Schneider 2011: xxv, 5, 16.
107 Breeze 1993: 17–18, 21–3.
108 Ibid.: 23; Wohl 1983: 244–5. Considering that Rawlinson and Way had already served in the Royal Commission on the Sewage of Towns, this is hardly surprising (ibid.).
109 Goddard 1996.
110 Ibid.: 275–85. For Berlin, see Mohajeri (2005: 154–67); for Paris see Reid (1991: 53–70).
111 Goddard 1996: 280, 284; Beder 1993.
112 Schneider 2011: 125–38.
113 Hamlin 1987: 274–6, 286–93.
114 Breeze 1993: 79.
115 Hamlin 1990: 172–4; Hamlin 1987: 381.
116 Breeze 1993: 29–30.
117 Luckin 1986: 163–9; Rosenthal 2014: 22–4.
118 'An Act for making further provision for the prevention of the pollution of rivers', in *Law Reports. The Public General Statutes, passed in the thirty-ninth and fortieth years of the reign of Her Majesty Queen Victoria, 1876, Vol. XI*, London, 1876, chapter 75, clauses 2–3.
119 Ibid., clause 6; Luckin: 1986: 170–1; Breeze 1993: 190.
120 'An Act for making further provision for the prevention of the pollution of rivers', clauses 4–6.
121 Luckin 1986: 165–8.
122 Rosenthal 2014; Hamlin 1988c; Wohl 1983: 249–56.
123 Wohl 1983: 256.

1   A Rivers Pollution Prevention Act for India?

In February 1890, the *British Medical Journal* called the attention of its readers to some disconcerting news it had received about recent events in Banaras. During Prince Albert Victor's visit to that city in January, the municipality had festively inaugurated its new water supply and sewerage projects, with the Prince laying the foundation stone for the waterworks. While appreciating the positive impact both projects were to have on the immediate cleanliness of Banaras and the health of its inhabitants, the article took strong exception to the proposed sewerage scheme:

> How are the mighty fallen! [...]. [T]he municipality intend[s] to discharge the sewage of the city into the Ganges a few miles below the town. As India has no enactment similar to the English Rivers Pollution Act there is no power to prevent their doing so, nor to prevent other cities in the Gangetic valley following their example, and thus converting the river by the time it reaches Calcutta into a gigantic cesspool [...]. [I]t is to be hoped that the Legislature will interfere before what is now only a possible danger of the distant future becomes a serious menace to all towns on the lower Ganges [...].[1]

The Banaras sewerage project, which in the eyes of the *Journal* and its Calcutta informant threatened to convert 'the noble river' into a 'sewer', and to reduce it 'to the condition of Father Thames',[2] was just one among several sewerage projects the North-Western Provinces were planning in the early 1890s. As part of his ambitious agenda on urban sanitary reform, Lieutenant-Governor Sir Auckland Colvin (1887–92) declared the introduction of water supplies and sewerage systems in his major cities a priority. Thus, projects similar to that in Banaras were envisaged for other riparian cities, namely Kanpur, Allahabad, Lucknow and Agra. Before we turn to some of these projects and the controversies they ignited, it is important to sketch the wider contemporary context of colonial policies on public health and urban sanitation in which they were situated.

Starting with the early 1860s, the Government of India developed a serious interest in sanitation, which was directly linked to the experience of the Great Rebellion of 1857/58. The large number of fatalities among British soldiers due not to combat, but disease, had distinctly brought to light the sanitary deficits within the military and their potential to jeopardise the stability of the Empire. A similar situation had occurred during the Crimean War a few years earlier, and both gave rise to a vigorous lobby in Britain demanding better health and sanitation within the British army. On these grounds, a Royal Commission on the Health of the Army in India was appointed in 1859. The commission's first report of 1863 dealt not only with directly military matters, such as the soldiers' diet and the living conditions in military barracks, but also with the sanitary state of Indian towns and populations in general. As the commission saw it, the health of the troops was inextricably connected with the health of the Indian people, especially when it came to the spread of epidemic diseases like cholera. As main sanitary defects it identified polluted water supplies, the absence of proper drainage and the general uncleanliness of the urban surroundings, which it blamed on the allegedly insanitary habits of the 'natives'.[3] In direct response to the report, the Government of India expanded its public health administration by establishing a sanitary branch within its home department and by appointing a sanitary commissioner. Additionally, sanitary branches were created in each province and provincial sanitary commissioners appointed. The newly created service complemented the Indian Medical Service (IMS), the existing medical establishment concerned with military and civilian health. As advisors to their governments without executive powers, the sanitary commissioners' duty was to inspect and report on the sanitary conditions in their provinces and to suggest measures for betterment. For this, they were to take regular tours and to collect vital and meteorological statistics. From the 1870s, they were moreover responsible for the spread of vaccination.[4]

As in Europe until the late nineteenth century, the miasmatic disease theory provided the intellectual backdrop to official public health policy in India, and of all diseases, it was cholera around which the greatest administrative concern revolved.[5] Fundamental in defining medical theory in India during the 1860s and 1870s was James L. Bryden, statistical officer to the newly formed sanitary branch and India's premier epidemiologist. According to Bryden, epidemic cholera was generated through the interaction of two processes: the reproduction and decay of pathogenic cholera 'seeds', which was accelerated or retarded by certain environmental conditions, and the epidemic spread of these 'seeds' beyond endemic areas. The latter, Bryden believed, was caused by monsoonal air currents, which accordingly determined the course and geographical reach of cholera epidemics. Combining this belief in the agency of specific meteorological conditions in determining epidemic zones, and the fact that India was the only country in which cholera was endemic, Bryden concluded that India was epidemiologically unique. While the spread of cholera in India followed a specific pattern defined by meteorological conditions, it might as well be spread by contagion or otherwise in Europe. The main object of sanitary policy according to Bryden was the removal of filth, the medium in which the cholera seed thrived, and the improvement of barracks and buildings, in order to prevent aerial incursions of the disease.[6]

The majority of medical officials in British India shared Bryden's views. Instrumental in translating his theoretical framework into practical measures was James McNabb Cuningham, the first sanitary commissioner to the Government of India from 1866 to 1884. His policy is aptly summed up in one of his last writings in office:

> Sanitary improvements, and sanitary improvements alone, embrace the whole action which a Government can take in order to prevent cholera. [...] Pure air, pure water, pure soil, good and sufficient food, proper clothing, and suitable healthy employment for both mind and body, these are the great requisites for resisting the cause or combination of causes which produces cholera.[7]

These official directives and their underlying concepts were challenged by a number of medical men who supported John Snow's contagionist, waterborne theory. The sanitary commissioner of the Punjab A.C.C. DeRenzy, for instance, repeatedly attacked government for its inactivity to prevent the spread of the disease and pressed for more specific measures to secure the quality of water supplies. Government answered mounting criticisms by resorting to Max von Pettenkofer, whose 'sub-soil water' theory included a hypothetical, communicable cholera germ, but stressed the primary importance of local environmental factors in its propagation. Where dissent became too strong, government was ready to take more rigorous action, as DeRenzy's case shows. Owing to his persistent criticisms of Bryden, he was ultimately transferred to military duties in a remote station in Assam.[8]

British Indian health policies also faced severe pressures from without. Between 1851 and 1894, eight international sanitary conferences were convened to address the threat of recurring cholera epidemics in Europe. There existed a wide consensus that cholera was somehow contagious, transmitted either directly from person to person or, indirectly, through contaminated food or water. Thus, delegates to the sanitary conferences demanded not only the imposition of quarantine regulations against ships coming from India, but also stricter measures to control the disease within the country itself.[9] The Government of India however remained adamant in its adherence to localist explanations of cholera. The reasons for this were economic as well as political. In the aftermath of the Great Rebellion, the colonial regime was wary of interfering with Indian religious practices by enforcing wider regulative measures, such as *cordons sanitaires* and quarantines during pilgrimages and religious fairs, apprehending that these would incite civil unrest.[10] Moreover, British trade by 1880 accounted for almost 80 per cent of the total tonnage passing through the Suez Canal. To accept that cholera was contagious and likely to be transmitted by polluted water, irrespective of geographic locality, would have forced government to accept quarantine regulations for ships coming from India as demanded by the international sanitary conferences. This would have considerably disturbed the flow of British trade and was therefore strongly resisted, not only in India but also in Britain itself.[11] Another reason for resistance was the institutional and intellectual rigidity that prevailed within the ranks of the IMS. As Mark Harrison has put it, the 'slowness of promotion [...], the pervasive anti-intellectualism, and bitter internal conflicts, fostered a climate in which innovation in theory and practice was positively discouraged'. Trends in medical theory and policy emanating from the metropole were therefore slowly responded to, or actively resisted.[12]

In the wake of Robert Koch's discovery of the *comma bacillus* in Calcutta in 1884, the Indian government geared up its defence. In the same year, it established the first medical laboratory in the capital and appointed the Scottish doctor David Douglas Cunningham as its director. While the laboratory marked the first step towards bacteriological research in India, Cunningham's agenda remained deeply entrenched in the political context. Over the next almost 30 years, he extensively researched and published on different aspects of cholera, producing the largest contribution from British India to the field. But rather than conducting original research, Cunningham's laboratory functioned as a tool with which the Indian government sought to disprove Koch's germ theory and legitimate its own sanitary policies. Closely aligned to Max von Pettenkofer's theories, Cunningham during the 1880s and 1890s acknowledged the existence of a cholera germ, but downplayed its role in causing the disease, insisting on the primacy of local conditions.[13]

It was only from the early 1890s that colonial medical theory and public health policies started to catch up with the mainstream of contemporary science. At the time, repeated waves of cholera, plague and kala-azar painfully exposed the helplessness of the existing medical establishment and its regime of sanitation, drugs and hospitals in effectively dealing with epidemic disease. Consequently, a growing number of British residents and medical officers demanded a fundamental reorientation of public health policies along the lines of new medical science, especially Pasteurian vaccine research. Within the IMS, the new enthusiasm for bacteriology and Pasteur emerged from a new generation of medical officers who were coming to serve in India from the 1880s. Many of them had received training in bacteriological methods in European research institutes, such as Koch's laboratory in Berlin and the Pasteur Institutes in France. Eventually, they launched a movement for the establishment of a Pasteur Institute in India, which bore fruit in the establishment of an anti-rabies Pasteur Institute at Kasauli in 1900.[14] Another reason for the Government of India's change of attitude was the fact that the long-feared economic threat inherent in accepting contagionist theories was getting weaker. While Britain had initially dominated shipping through the Suez Canal, other European powers increased their own traffic considerably from the 1880s. Moreover, even Robert Koch argued against quarantine regulations, since he believed that cholera would reach Europe anyway over land. Both led to the liberalisation of quarantine regulations in the 1890s.[15]

On the background of these developments, germ theory could no longer be completely denied by the mid-1890s, even in British India.[16] The growing official rapprochement towards bacteriology found expression on an institutional level, too, with the opening of a bacteriological department at the Poona College of Science in early 1891 under Alfred Lingard, who was designated as 'Imperial Bacteriologist'. In 1892, the North-Western Provinces called the British bacteriologist Ernest Hanbury Hankin to take charge of their newly established government laboratory at Agra, which was to take care of much of Northern India. One year later, Waldemar Haffkine was sent to India by Louis Pasteur to test his cholera vaccine, which proved an instant success. In 1896, Haffkine set up a plague research laboratory in Bombay under the control of the Bombay Government, the tasks of which comprised the preparation of anti-plague serums and anti-cholera vaccinations, and the examination of pathological specimens, among other things.[17] Thus, germ theory and Pasteurism had come to be widely recognised in British India by the turn of the twentieth century.

A major stage for colonial sanitary policy to unfold was the city. During the second half of the nineteenth century, urban sanitation in general and the construction of water supply, drainage and sewerage infrastructures in specific were moored in two overarching colonial discourses. For one, they were part of the colonial agenda concerned with the maintenance of power after the Great Rebellion. After 1858, Delhi, Lucknow and other North Indian cities that had acted as hotbeds of insurgency were reshaped along three principles: safety, sanitation and loyalty. First, urban morphologies were changed in such a way as to guarantee military safety, making the city easy to defend. From this emerged the pattern of the colonial city, in which newly built European cantonments and civil sections were set against old, neglected quarters inhabited by the 'native' population. Second, sanitation was promoted in order to protect the health of European soldiers and civilians. The priority being European health, sanitary infrastructures were primarily introduced in European quarters. And third, the loyalty of the Indian citizen was to be secured by fostering the loyalty of Indian elites.[18]

The second colonial discourse that encompassed urban sanitation was that of progress and modernity. This discourse had been prominently developed under the utilitarian government of Governor-General William Bentinck (1828–35), and was moulded specifically around the introduction of Western technologies in India under Governor-General Lord Dalhousie (1848–56). Thus, Dalhousie's tenure saw the construction of the Indian railways as well as the introduction of the telegraph.[19] In the urban context, the replacement of traditional sources of water supply, such as wells and cisterns, with a centralised piped water supply, and the introduction of sewerage systems in place of dry conservancy was viewed as an integral part of urban modernisation.[20] After 1858, the discourse on progress and modernity itself was closely connected to strategies of power maintenance. While the Crown refrained from overt interventions into Indian religious and cultural life in fear of creating civil unrest, social control was exerted all the more strongly, albeit indirectly, on the municipal level, by directing individual and social behaviour through the enforcement of sanitation and building regulations.[21]

The first steps towards improved urban sanitation, even before the 'sanitary awakening' of the early 1860s, were made in Bombay and Calcutta, both port cities of great economic and political importance with a large European population. In Calcutta, a combined system of storm water drainage and sewerage was proposed in 1855 and operative from 1868 onwards, and in 1870, filtered water from the Hooghly began to be supplied through the new waterworks at Palta. In Bombay, too, waterworks were inaugurated in 1859, while drainage and sewerage systems for a long time remained rudimentary due to financial constraints and opposition from within the municipality. The know-how and material for these urban technologies were imported from Britain, and were almost identical to those used in European cities.[22] Beyond the centres of trade and power, urban sanitation developed at a much slower pace. The sanitary movement materialised at a moment when the Government of India was actually keen to free itself of financial responsibilities. Thus, while it promoted urban sanitary improvement, it was eager to devolve the costs for it to the municipal governments.[23] Municipal bodies could be formally instituted from the 1840s. Under the provisions of Act XXVI of 1850, municipal functions included conservancy, road repairs, lighting, the framing of by-laws and their enforcement, as well as powers of taxation. However, the establishment of a municipality depended on the wish of the inhabitants. On the background of the ever increasing 'India debt' (standing at 98 million pounds after suppression of the Great Rebellion), the Indian government from the 1860s undertook several steps towards the decentralisation of its fiscal system in an attempt to relieve imperial finance. Draft proposals of 1861 included the transfer of financial responsibilities for roads and public works to local bodies, but the details were left in the hands of the provincial governments. In 1864, a resolution by Governor-General Lord Lawrence promoting local self-government transferred the responsibility for police charges to the municipalities, who in turn were allowed to levy taxes to cover the expenses. Following the resolution, legislation was enacted in almost every major province and by the end of the 1860s, every Indian town of importance had become a municipality. In 1870, Governor-General Lord Mayo transferred the administrative and financial responsibility for sanitation, road maintenance and other local services from the imperial to the provincial governments, who in turn passed sanitary expenditure on to the municipalities. As a concession to the Indian taxpayer, who was to shoulder the financial burden of sanitary expenditure, the representative element in municipal bodies was reintroduced. This element was further enlarged under the liberal government of Governor-General Lord Ripon in 1882.[24]

How much municipalities spent on sanitation varied greatly and largely depended on the sources of municipal income. Fixed taxation (e.g. house and land taxes) provided a more steady income than economic revenues (e.g. from octroi and terminal taxes) and made it easier to budget for sanitary expenditure. The budget was further determined by the willingness of provincial governments to grant loans or grants-in-aid for sanitary improvements. Another important factor, especially after 1882, was the attitude of Indian ratepayers towards western concepts of public health.[25] In this context, large-scale infrastructure projects like waterworks and sewerage systems lay beyond what most municipalities could and wanted to afford, and urban sanitation remained predominantly confined to less cost-intensive measures. Among these stood improved drainage and general cleanliness, the construction of latrines, the relocation of 'nuisant' trades, and the control of interments.[26]

Under the government of Governor-General Lord Dufferin (1884–88), the grounds for urban sanitary reform in India became more conducive. Keen to promote public health, Dufferin in 1887 put forward a set of administrative innovations to be implemented at the provincial level. His resolution directed each province to (i) set up a central authority to advise, direct, supervise and control sanitary projects; (ii) promote facilities for scientific research and experiment for the investigation of disease, and (iii) promote sanitary education in schools. The Government of India in turn would provide loans to provincial governments at moderate rates for sanitary projects of more than local importance. In fulfilment of the first point, Sanitary Boards were established in each province. Primarily concerned with large sanitary engineering projects, these were to act as advisory bodies to municipalities on sanitation and to authorise their plans for sanitary projects.[27]

In the North-Western Provinces, Dufferin's resolution stirred an unprecedented move towards major urban sanitary infrastructures. During Sir Auckland Colvin's tenure, all the great cities built their first waterworks: Agra's were operational from 1890, Allahabad's from 1891, Banaras's from 1892, and those of Kanpur and Lucknow from 1894. Between 1890 and 1894, the annual loans granted to municipalities of the provinces for such projects ranged from 10 to 15 lakhs of rupees.[28]

The introduction of centralised water supplies raised several questions: What kind of system should be built to remove the great volumes of wastewater thus created? How should the wastewater be disposed of? And should excreta continue to be collected and removed manually, or should they be put into the sewers instead? Since all the major cities of the North-Western Provinces lay along rivers – Kanpur, Allahabad and Banaras along the Ganges, Lucknow along the Gomti, and Agra along the Yamuna – the easiest way to get rid of sewage was to direct it into these rivers. This, however, turned out to be a disputed approach, as the Banaras controversy was to show.

Banaras in the nineteenth and early twentieth centuries was both the largest urban centre of the eastern Gangetic plain, counting a resident population of about 200,000 people, and a major site for Hindu pilgrimage (see Figure 1.1).[29] Like other pilgrimage sites, it attracted thousands of visitors all over the year, and many more during great religious festivals. While such big gatherings of people naturally put enormous pressure on municipal infrastructures, they also offered an ideal nurturing ground for disease. Haridwar and Allahabad, for instance, saw repeated major cholera outbreaks during the Kumbh Mela.[30] Banaras, with its location between the most important pilgrimage places to the East (e.g. Puri, Baidyanath and Gaya) and the West (e.g. Allahabad, Haridwar and the pilgrimage sites in the Western Himalayas) was a critical junction within the 'northern epidemic highway', over which cholera and other diseases rapidly spread.[31]

Figure 1.1 Banaras (source: K. Baedeker (1914) *Indien: Handbuch für Reisende.* Leipzig: Verlag von Karl Baedeker).

The religious significance of the city was and is defined by its inherent status within the sacred geography of Hinduism, as well as by its location on the banks of the Ganges. According to Hindu scriptures, the river Varana to the North and the rivulet Assi to the South define the city as a *tīrtha*, a sacred zone that acts as a doorway between heaven and earth. The souls of those who die within this sacred zone attain the blessings of the god Shiva and are liberated from the cycle of reincarnation. The presence of the Ganges, the intermediary between this world and the heavenly realms, adds to the significance of the *tīrtha* and makes it most desirable for the faithful to die at Banaras, or to have their ashes strewn into the sacred river there.[32]

As a major urban centre and pilgrimage site with the reputation for being a hub of epidemic disease, Banaras offered itself to the scrutiny of the colonial government. In the eyes of British observers, the sanitary condition of Banaras posed an immediate threat to the city and its surroundings. In his *Preliminary Report on the Sewerage and Water Supply of the City of Benaras* of 1880, super-intending engineer Frank Fitzjames wrote:

> When I state that it may be said to be without drainage of any kind, that its subsoil is saturated to a depth of several feet with the filth and abominations of centuries, that every well in the city is contaminated by percolation from the offensive soil, and that in spite of all these evils it is a healthy city, one wonders how this can be. [...] The Benaras stinks [...] only too plainly and emphatically tell us how much drainage is wanted in Benaras and there cannot be much doubt that if this city is not very thoroughly drained and improved, as much as sanitary science will allow, an outbreak of some epi-demic like the plague will devastate the city, and cause heavy loss to Government.[33]

Fitzjames was given the task of designing a comprehensive water supply and sewerage system for Banaras, which was to be the first of its kind in the North-Western Provinces. Much as the city itself, the Ganges and the *ghāṭs*[34] along its banks appeared in a bad light during the inspections. In the same year, the sanitary commissioner of the North-Western Provinces, Charles Planck, noted:

> Continuing to the line of bathing-ghats, world famous I believe, [...] I notice small house drains from above, pouring their contents, in slowly trickling stream, alongside and even over the steps of ghats in several places, the streams ending in and mingling with the Ganges water, in which persons in thousands are now bathing and several water-carriers dipping their pitchers. [...] Arrived at the mouth of the great Dasasumedh sewer, [...] I see that it just fails to reach the water at a place where many people are landing from boats. The sewer pours out a stream of black-coloured sewage about six inches deep, which [...] serves to darken the water for full 50 yards in length along the shore. The boats, with their freights of passengers, come from above and across the river to join in the ceremonies of the day[,] lie in this dark water[,] and the passengers land with difficulty and complaint because of the black mud and powerful stench to which the sewer gives origin.[35]

The drains Planck mentions were laid in the 1790s and the 1820s, and designed to lead domestic wastewater (including excreta) from riparian houses over the *ghāṭs* into the Ganges.[36] During the rainy season, these drains also received the overflowing contents of the city ditches.[37] Fitzjames's' project envisaged the construction of a large underground sewer, which would intercept the numerous smaller drains carrying the wastewater from city households, and discharge its contents into the Varana, shortly before its confluence with the Ganges.[38]

No immediate action appears to have been taken with regard to Fitzjames's' proposals. But the discourse about the sanitary state of the city and its river had stirred the interest of the local Indian elite. On the initiative of the *dīvān* (minister) of the Maharaja of Banaras, Diwan Ramchand, the newly founded Kashi Ganga Prasadini Sabha[39] held its inaugural meeting in the town hall on 19 November 1886. Earlier that year, the *dīvān* and Raja Shivaprasad[40] had approached the sanitary commissioner with a desire to point out the main sanitary defects of the city, which in their eyes urgently needed to be remedied: the polluted condition of the Ganges and the river banks, and that of the city ditches and the minor drains in the city centre.[41] The proclaimed aim of the Sabha was 'to concert measures to prevent the pollution of the Ganges, within the limits of the sacred city of Benares, by the discharge of filth into it through the sewers'.[42] The leading members of the Sabha included the Maharaja of Banaras, Diwan Ramchand, Raja Shivaprasad and other Hindu men from the nobility and influential sections of society. During the inaugural meeting, sub-scriptions for the considerable sum of 250,000 rupees were obtained, with which the Sabha intended to put its aims into practice. Most importantly, the money was meant to relieve the poorer classes from the burden of increased taxation, which a large sewerage project would certainly entail.[43]

The foundation of the Sabha was warmly welcomed by British officials. Sanitary commissioner Planck earlier had already expressed his happiness about the 'considerable awakening, in regard to the advantages and necessity of sanitary improvement' shown by the Banarasi elite.[44] Sir Auckland Colvin was likewise pleased, although this appears to have been motivated mostly by the Sabha's projected financial contribution. As a sign of support, the Lieutenant-Governor accepted the Sabha's invitation to preside over it as a patron.[45]

With the local elite taking active interest, government entrusted its engineer A.J. Hughes with the revision of Fitzjames' earlier plans. Hughes had been selected as Superintending Engineer for Municipal Works in 1888, shortly before the introduction of the provincial Sanitary Board, to advise municipalities with an interest in water supply and drainage, prepare respective schemes and supervise their execution.[46] Like Fitzjames, Hughes proposed the construction of a large underground sewer, intercepting sewage from an extensive network of small sewers and discharging it into the Ganges, shortly before its confluence with the Varana. Householders keen to connect their latrines and houses directly to the sewers could apply for a house connection. Excreta and other wastes from households not so connected would be cleared by sweepers and disposed of in so-called 'pail depots', i.e. collection points linked to the underground sewer through a pipe.[47] As to the outfall of the main sewer, Hughes proposed the instal-lation of a sewage farm once such a project was financially viable, and once the quantity and quality of the sewage was known.[48] In January 1890, the water-works and sewerage projects were inaugurated with great fanfare.[49]

Shortly after the day of inauguration, the Government of India received a letter of protest by the sanitary commissioner of Bengal, W.H. Gregg, in which he vehemently opposed the discharge of large amounts of untreated sewage into the Ganges. 'There can be no harm', wrote Gregg, 'from surface drainage being discharged into the Ganges, but if that river is to be made a general sewer for towns on its bank, the ill-effects of such a scheme [...] cannot be exaggerated'. To do so would 'seriously affect the health of the people living in the towns situated on the banks of the river below Benares who obtain their supply of drinking water from the Ganges'.[50] According to Gregg, the Banaras project brought to light the urgent need of a national law to prevent municipalities from 'using the Ganges as a sewer', and presented a 'favourable opportunity' to pass a corresponding Act.[51] Gregg's protest was supported by the Lieutenant-Governor of Bengal, who found the issue to be of great importance.[52]

Confronted with Bengal's protest through the Government of India, Sir Auckland Colvin asked the latter to lay out general principles with regard to the discharge of 'town drainage' into rivers, should there be any objections to the continuance and extension of this practice. These principles, the Lieutenant-Governor suggested, should be set up after full and exhaustive enquiry, and without special regard to the case of Banaras.[53] Thus, faced with demands from both Bengal and the North-Western Provinces, the Indian government was moved to consider a national law, or at least national guidelines, on the disposal of sewage into Indian rivers.

In the absence of any large-scale pollution of Indian rivers, no national river pollution law existed or had been discussed up to the late nineteenth century.[54] Laws pertaining to rivers, such as the Obstruction in Fairways Act of 1881, were mainly concerned with navigability, prohibiting encroachments and other obstacles that may impede shipping. Earlier provincial Acts included similar provisions, such as the Shore Nuisance (Bombay and Kolaba) Act of 1853.[55] Water quality was addressed on a general level by the Indian Penal Code of 1860. Section 277 held that 'whoever voluntarily corrupts or fouls the water of any public spring or reservoir, so as to render it less fit for the purpose for which it is ordinarily used' would be liable to imprisonment or a fine.[56] The Penal Code has been described as a landmark in the Indian history of the legal control of environmental pollution, as it contained several provisions aimed at the control of water and air pollution.[57] Thus, section 278 prescribed punishment for 'whoever voluntarily vitiates the atmosphere in any place so as to make it noxious to the health of persons', and section 430 on a more general level prohibited any acts which would diminish available water supplies used for agriculture, drinking, etc. River pollution could also be tried as a 'public nuisance' under section 290, which was defined as 'an act or [...] legal omission, which causes any common injury, obstruction, danger or annoyance to persons'.[58] The enforcement of these provisions however was problematic, as the Code left the burden of proof with the prosecutor.[59] Additionally, the definition of 'public spring or reservoir' in section 277 did not include any direct reference to rivers, which is why the cases tried under this section were almost exclusively concerned with the pollution of well water.[60] Similar clauses aimed at the protection of water sources from pollution were included in a number of provincial laws, such as the Northern India Canal and Drainage Act of 1873, which provided imprisonment or a fine for anyone who ' "corrupt[ed] or foul[ed]" the water of any canal so as to render it less fit for the purposes for which it [was] ordinarily used'.[61] While these state and provincial laws certainly reflect a growing concern for the protection of water sources from pollution, they were seldom applied and overall remained dead letter.[62]

The most prominent issue for various governments up to the late nineteenth century was the disposal of uncremated or insufficiently cremated human corpses into rivers, especially into the Ganges. Official complaints against this practice reached a peak during times of excessive mortality caused by famines or epidemics, when only the richer segment of Hindu society was able to afford the wood needed for the cremation of their deceased. While these complaints were certainly rooted in notions of sanitation to some extent, they contained a strong aesthetic and moral element that predominated over the former. On the occasion of the Great Famine of 1876–78 the sanitary commissioner of the North-Western Provinces vividly described the bloated corpses floating in the Ganges, with stretched out limbs or half eaten by fish, jackals and the like, and complained of the immorality of this practice, so fraught with 'disrespect for the dead'. In his view, 'of all things Indian, probably nothing so earnestly calls for reform, as this question of the disposal of the dead amongst the Hindus'.[63] Similar complaints from health officials more outspokenly raised the question of regulation. The sanitary commissioner of Madras one year earlier had suggested 'on public health grounds' that an Act might be passed prohibiting the disposal of bodies into rivers, as well as interments in river beds and on river banks.[64] However, the Indian government was not inclined to pass any legislation,[65] possibly because of its reluctance to directly interfere with Indian religious practices.

On the municipal level, too, the issue was generally treated with care. The only municipality that effectively passed a law was Calcutta. As a consequence of the fever epidemic ravaging several districts of Bengal in 1862/63, the number of uncremated corpses floating in the Hooghly at Calcutta increased and became a positive nuisance in the eyes of the Bengal government. To make things worse, the *murdā farrāśes* employed with sinking the corpses struck work, demanding higher wages. Following requests from the Bengal government, the Justices of the Peace of Calcutta in 1864 passed a by-law prohibiting the disposal of corpses and animal carcasses into any river, rivulet or canal within the town limits of Calcutta.[66] The law was part of a more general effort to keep the Hooghly free from pollution; soon after, arrangements were made to transport the night soil (hitherto collected and dumped into the river) beyond city limits for trenching.[67] While the by-law enjoyed considerable support among the Western-educated Indian classes, it seems to have caused great anxiety among the British residents of Calcutta. Invoking the experience of the Great Rebellion, the *Englishman* severely criticised the government and admonished it of the incalculable reaction of the Hindu masses.[68]

In 1890 then, laws applicable to river pollution were mostly part of a very general legal framework to keep water sources and urban environments clean, or, in the case of Calcutta, aimed at tackling a specific pollution problem. The Banaras sewerage project acted as a turning point, igniting the first debate about a national river pollution law. It did so because it added a completely new dimension to the question of river pollution in India. For one, it envisioned the organised disposal of the sewage from 200,000 people into an inland river, something which had never been done before. Calcutta did dispose of its sewage into the Bidyahari from 1868 onwards, but this river discharged more or less directly into the sea. Second, the Banaras project was proposed at a critical moment when germ theory had come to seriously challenge miasmatic theories in Europe and had slowly started to gain momentum in British India. While under the miasmatic model, rivers appeared as viable waste carriers, germ theory introduced a completely new perspective and raised urgent questions about the dangers of sewage disposal into rivers.

Bengal's request for a national river pollution law met with little enthusiasm from the Indian government. The only official to lend his strong support to W.H. Gregg was the Sanitary Commissioner of India, W.R. Rice. To Rice, the provisions of British law (namely, the Rivers Pollution Act) offered the best guiding principle in a matter with which India had no experience so far. These provisions were validated by the recommendations of the Royal Commissions on Town Sewage and Rivers Pollution, and by the opinions of some of the most eminent British sanitary experts of the time, such as Edmund Parkes and Baldwin Latham. Both warned of the threat to public health caused by the discharge of untreated sewage into inland streams, and strongly advocated its prior treatment on sewage farms.[69] Hence, Rice advised that Indian towns should be prohibited from discharging untreated sewage into rivers and be committed to treat it either by a chemical method or by applying it on sewage farms. In accordance with his sources, Rice also found the theory on the self-purification of rivers highly doubtful. Under Indian conditions, where rivers were 'wide, slowly-running streams', often breaking up into 'back water and stagnant pools' during the dry season, efficient self-purification seemed to be even more unlikely, as rivers did not provide the swift running of water assumed to be a prerequisite for the process. Another factor which, in the eyes of the sanitary commissioner, increased the danger to public health, was that riparian populations in India (other than in Britain) were in the habit of drawing water for drinking and domestic purposes directly from rivers without filtration, and were therefore directly affected by its quality.[70]

Officials in the home department's sanitary branch, where the issue was first debated, were less convinced about the feasibility of British law for India. Reversing Rice's argument, one member thought the very different character of Indian rivers, the beds of which were 'swept clean every monsoon', to be the best safeguard against harmful effects of sewage on water quality. Apprehensions were also expressed with regard to finance, as public works built for the prevention of river pollution were bound to entail great costs. However, Rice's argument about the dangers of a potentially polluted, unfiltered water supply carried great weight. Well aware that the matter in question not only touched upon Banaras, but also numerous other sewerage projects planned by the North-Western Provinces, the sanitary branch agreed on the necessity of general guidelines, and called for a special committee of inquiry.[71]

While Bengal's intervention thus met with some initial support, the Legislative Council of India unanimously rejected it. British law, according to Governor-General Lansdowne, was 'quite inapplicable', and contained provisions (not further specified) which could not be enforced in India. Moreover, any committee set up to devise general guidelines would face a 'task of portentous dimensions'. Rather, a 'common-sense scheme' should be adapted in each case, taking into account the specific circumstances of each city.[72]

Lansdowne's opinion was closely followed by the council members. Gregg's argument, they agreed, was vague and lacked verification, and so did not justify an extensive inquiry. Moreover, it was pointed out that streams rapidly purified themselves from sewage, a process that was thought to be enhanced by the volume and length of Indian rivers. This argument was based on personal observation (for example of the Shamal stream near Shimla), and on certain unnamed scientific reports. Consequently, most of the members found the issue of no importance in the Indian context, or even, as one member put it, downright 'absurd'. However, the council knew that Bengal's objections could not be simply ignored. They therefore agreed that, following Rice's advice, the relevance of the issue should be tested empirically, by conducting chemical, biological and 'microscopical' tests of river water above and below Banaras. On the basis of these tests, the extent to which the Ganges was being polluted by the sewage entering it through the old, existing drains could be assessed.[73]

Summarising these deliberations, the Indian government informed the North-Western Provinces that 'English law on the subject [cannot], without much further enquiry and consideration, be applied to the conditions of India', and that government therefore was not ready to lay down any 'general principles' on the discharge of sewage into Indian rivers. But as it viewed the matter as one 'of the first importance in its bearing upon the sanitation of Indian towns', the North-Western Provinces were directed to

> make enquiries into the matter in any localities where town sewage is already conducted into rivers [...] and ascertain by actual testing and experiment whether the effect upon the water of the rivers is such as to render it dangerous to public health.

In the meantime, it gave Sir Auckland Colvin a free hand to proceed with the Banaras project as he thought fit.[74]

The Legislative Council's verdict temporarily transferred the debate to the level of provincial government. However, as critical voices had anticipated, it was bound to resurface once other sewerage projects stood ready for execution, and it did so just around two years later in relation to Kanpur.

The city of Kanpur (see Figure 1.2) in 1891 had a population of roughly 190,000, and with its large cantonment area and government ordnance factories was a place of major military importance.[75] As such, its sanitary state was of great official concern. In 1872, the city built its first surface drainage system. The existing two natural drainage lines, small streams it may be assumed, were

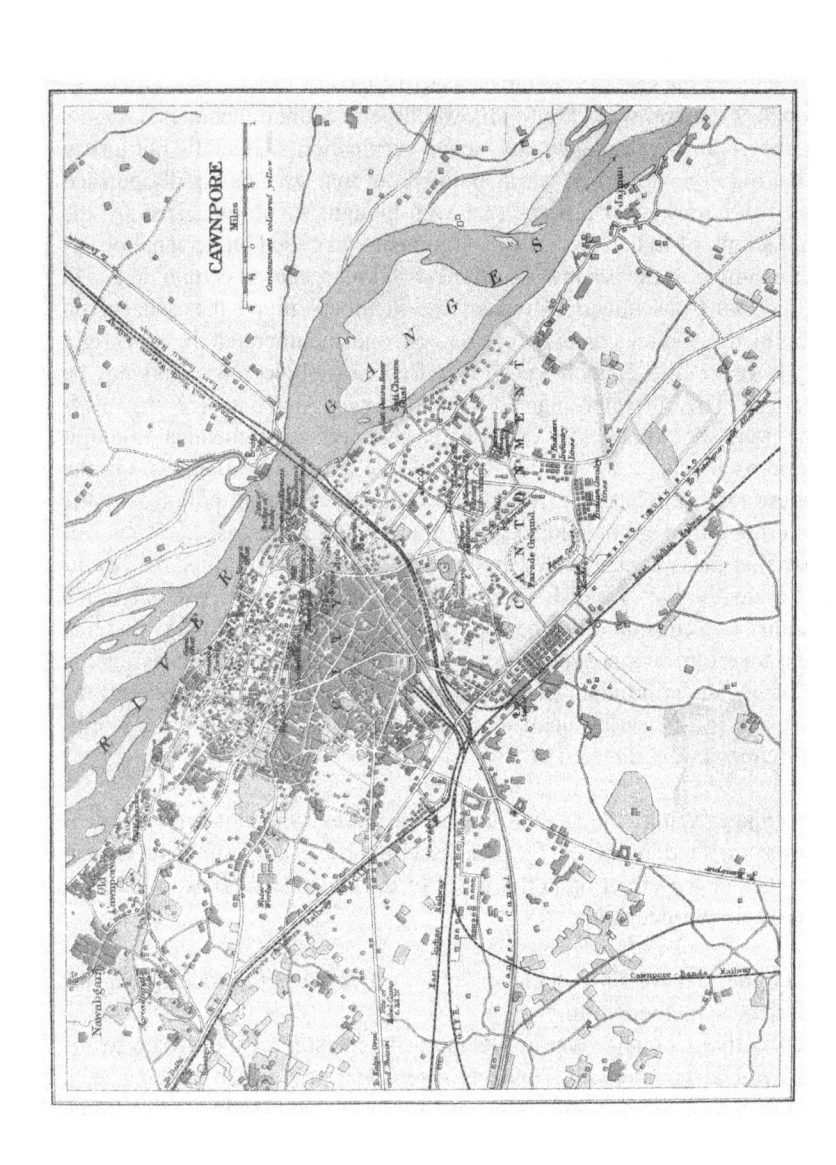

Figure 1.2 Kanpur (source: J. Murray (1933) *A Handbook for Travellers in India, Burma and Ceylon, including all British India, the Portuguese and French Possessions and the Indian States*. 14th edn, London: Murray).

converted into main sewers (called the Sisamau and Jajmau sewers), being lined with masonry and partly covered. After receiving the contents of a network of smaller surface drains, the sewers emptied into the Ganges right above the principal bathing *ghāṭs* and the cantonment. Over the years, the system had proved largely inefficient due to deficits in material and construction. Excreta were still removed by hand and cart to trenching sites, but a great amount, together with other kinds of waste, found its way into the surface drains. Inefficient drainage and excreta removal, as well as polluted water supplies, were blamed to be the main cause for Kanpur's insanitary conditions and its extraordinary mortality rate of 50 to 65 per thousand, one of the highest in the North-Western Provinces. In the early 1890s, a combined water supply and sewerage project was devised to remedy the situation. The project largely followed the Banaras model: waterworks were to supply filtered water from the Ganges, and the existing surface drains were to be replaced with underground sewers, the latter designed to receive all the excreta through pail depots and at a later stage through direct house connections. As per the outfall, it was planned that the main intercepting sewer would initially discharge the sewage untreated into the Ganges. As soon as the finances of the municipality allowed it, a sewage farm or other purification measures would be added. As a preparatory arrangement for the latter, the sewage outfall had to be placed further down the city, where suitable land for a sewage farm was available. This meant that the main intercepting sewer had to be taken all the way underground through the military cantonment.[76]

In July 1892, India's military department – independently from the home department – raised objections against the running of the proposed outfall sewer underneath the cantonment. The atmosphere of the cantonment, the military authorities apprehended, would be 'fouled' by 'effluvia' emanating from deposits of faecal matter within the sewer. At the same time, the department objected to the present discharge of sewage into the Ganges above the cantonment. Instead, it suggested the application of the hand removal system (including a tramway), i.e. the collection of the sewage and its disposal as fertiliser on a sewage farm.[77] The military department thus opposed the sewerage project in two principal points: the means of transporting the sewage and the site of its ultimate disposal. Yet again confronted with opposition to one of their sewerage projects, the North-Western Provinces turned to India's home department and, pointing to the discord within the Government of India about the question of sewage disposal, reinforced their request for definite guidelines.[78]

Accepting that the issue could no longer be deferred, the Indian government finally agreed to look into the question and set up a committee of inquiry in early 1893. At the same time, however, several officials involved in the discussion were doubtful whether general guidelines were at all feasible, and to what extent the proposed inquiry could be useful.[79]

The committee was headed by the sanitary commissioner of India, W.R. Rice, and was to look into three questions, referring both to the general case and the specific case of Kanpur: (i) whether solid excreta and sewage in India were to be disposed of by dry carriage or water carriage in sewers, (ii) whether and under what conditions sewers were allowed to discharge into rivers, and (iii) whether or not it was practical to operate sewage farms. As the North-Western Provinces had not submitted any results of their water analyses yet, the information at hand to the committee was much the same as it had been in 1890.

Unfortunately, the committee's detailed discussions don't seem to have been documented, but in its final report, it stipulated the following guidelines: (i) where water was abundant, water carriage was preferable for the removal of both solid excreta and sewage, (ii) untreated sewage 'should not on any account be discharged into streams or rivers, but should be purified by filtration or otherwise prior to discharge', and (iii) where possible, sewage of inland cities and towns should be discharged upon sewage farms. In accordance with these guidelines, the committee directed that the sewage of Kanpur must be disposed of by water carriage onto a sewage farm. This was to be built to the East of the cantonment, where there was suitable, unsaturated land, the sewer running underground through the cantonment.[80]

What had started as a debate around a national river pollution law in 1890 came down to a committee of inquiry issuing general guidelines in 1893. These guidelines, however, hardly represented an official consensus; rather, they reflected the departmental and personal viewpoints of its members and the influential position of the sanitary commissioner of India. Headed by W.R. Rice, the committee was made up of four officials, two each nominated by the North-Western Provinces and India's military department, respectively. The sanitary commissioner of the former, G. Hutcheson, was outspokenly opposed to the disposal of untreated sewage into rivers, even if he had come to accept it as a temporary measure.[81] Moreover, both he and the military department shared Rice's strong bias for sewage farms. The only compromise was made by the military department, which stepped down from its initial resistance against the water carriage system.

In laying down its guidelines, the committee ignored the wide divergence that marked official opinion on the disposal of sewage into rivers in the early 1890s. The main dividing principle was the question if and how British law and expertise were applicable to Indian conditions, a question which could be answered in a variety of ways. The sanitary commissioner of India viewed British law and expertise as the most authoritative source for guidance, but his reference to the reports of the Rivers Pollution Commission and others was highly selective and could be challenged with numerous other 'reports' and expert opinions. The majority of officials, in contrast, based their opinion on a colonial ideology of difference.[82] From this perspective, due to their length and volume, Indian rivers appeared as suitable, self-purifying receptacles for sewage, which made the question raised by Bengal appear irrelevant. This ideology of difference dovetailed with the colonial paradigm of 'tropicality', which categorised the Indian environment as 'tropical', and as such as inherently different from 'temperate' or 'moderate' European environments.[83] As we shall see, the same set of arguments was spelt out very prominently, and in a much more elaborate manner, during the extensive discussions led within the government of the North-Western Provinces.

A most striking feature of the debate is the conspicuous absence of any argument about the precise implications of sewage-polluted river water for the spread of epidemic disease. W.H. Gregg's protest itself made no reference to the exact nature of the possible threat to public health, even though elsewhere he clearly expressed his conviction that cholera was principally transmitted by the consumption of polluted water, and that the faecal discharges of cholera patients contained the 'germs of the disease'.[84] This peculiar silence around germ theory, I suggest, reflects the delicate position it still occupied within the highest official circles in British India during the early 1890s.

The first debate around a national Indian river pollution law fits well with the larger historical context of colonial 'resource conservation' policy and related legislation in British India. In view of the lively scholarly debate that has revolved around this subject for over two decades, it will be valuable to look at our debate from this broader historical perspective.[85] While colonial governance in British India was on the whole characterised by a laissez-faire doctrine and non-interventionist policy, it is possible to make out two main trends of 'environmental governmentality' during the nineteenth and early twentieth centuries. First, municipalities tried to control and regulate urban environments in an effort to improve urban sanitation. Second, the Government of India sought to counteract the overexploitation of natural resources, such as forests, wildlife and fisheries, by regulation and protection.[86] British rule in India and elsewhere, with its unprecedented demands for raw materials, commodities, crops and minerals, created enormous pressures on the natural environment. Already during Company rule, the environmental consequences brought about by these intensive levels of extraction – especially deforestation with attendant soil desiccation, and soil salinisation due to large-scale irrigation – caused alarm and triggered a range of initiatives to contain environmental degradation and resource depletion.[87] An exemplary case for the Indian government's environmental interventionism is the Forest Act of 1878. The construction and expansion of the Indian railways from the 1850s exacerbated deforestation, timber being required as construction material, fuelwood and raw material for the fabrication of sleepers. In order to check deforestation, the Government of India in 1864 launched systematic forest management under the newly established Indian Forest Department. As a legislative backing to the new department, a first Forest Act was hurriedly drafted one year later, which enabled the state to legally acquire forest areas most urgently needed for timber supplies.[88] While the first Forest Act did not interfere with existing user rights, the second Forest Act of 1878 turned vast stretches of forests into state property and functioned as a legal tool to restrict access to local populations who had customarily depended on woodlands and forest products for their livelihood. By the end of the twentieth century, one fifth of India's forests were marked as 'reserved' forest, where user rights lay entirely with the state. This exclusivist policy triggered continual and widespread popular resistance until the end of colonial rule.[89] Other natural resources also became objects of state concern in the course of the late nineteenth century. In 1897, after almost two decades of debate, the Indian Fisheries Act was passed over the fast depletion of fisheries caused by overfishing and destructive fishing practices. It outlawed the use of dynamite and poison, but left it to the provinces to pass further regulations, e.g. on the dimension of fishing nets.[90] From the 1880s, after decades of state-induced slaughter of tigers, elephants and other so-called 'vermin' of the forest that were seen to encroach upon farmers' crops and live-stock – and thus upon colonial revenues – several provinces initiated first steps towards the protection of certain species under the existing forest laws. Starting with limitations on 'bag limits' for sport hunters, these soon included more far-reaching measures, such as wildlife reserves. However, the Indian government refrained from passing a general Game Law, wary of adverse reactions from cultivators.[91]

As the forest, fisheries and wildlife laws and regulations of the late nineteenth century illustrate, imperial 'environmental governmentality' in British India was marked by a tension between exploitation and conservation, a tension which was characteristic for the whole of the British Empire.[92] In the context of Indian forest history, environmental historians have long discussed the motives behind state 'conservationism'. Setting the tone for many later works, Madhav Gadgil and Ramachandra Guha in one of the earliest studies on the subject have argued that the primary incentive to colonial forest policies was financial and strategic, aimed at the continuous supply of timber and the creation of revenues.[93] In con-trast, Richard Grove in his seminal work *Green Imperialism* contended that state regulation was rooted in a genuine conservationist perspective, brought forward by a group of Scottish-educated doctors/scientists serving under the East India Company. These officials believed that the rapid deforestation on the subconti-nent was directly responsible for desiccation, soil erosion and climatic change, and thus for the increase in famines and droughts. According to Grove, it was their continual lobbying for conservation that led to the institutionalisation of state forest management.[94] As Mahesh Rangarajan has noted, the difference between these two seemingly contradicting perspectives may be explained with a difference in chronological focus. Whereas Grove concentrates on early nine-teenth century officials and their broader social concerns, Gadgil, Guha and others have looked at the production-centred agenda characteristic of a later gen-eration of officials.[95] More recently, authors have essentially supported Grove's theses, while criticising certain aspects.[96] The historiographies on colonial fisher-ies and wildlife policies similarly point towards a mix of aspirations lying at the roots of state regulation, including financial, conservationist and other motives.[97]

As pointed out by several scholars, state policies and laws on natural resources did not necessarily reflect any broad official consensus, but were often only passed after heated debates within the administration. The Forest Act of 1878 essentially represented the position of what Ramachandra Guha has called the 'annexationist' group, who aimed at total state control over all forest areas. While they were ultimately successful, their position was repeatedly challenged by the 'populists', who held that tribals and peasants must retain sovereign rights over woodlands, and the 'pragmatics', who sought to restrict state management to ecologically sensitive and strategically valuable forests.[98] The Indian Fisheries

Act of 1897 was equally controversial. The Inspector-General of Fisheries, Major Surgeon Francis Day, was the first to raise the alarm about the 'wasteful destruction' of fisheries in the 1870s. In a report to the Indian government, Day warned that overfishing and the destruction of breeding fish and young fish fry was causing a rapid decrease in freshwater fish, which he blamed on the state's failure to regulate access to fisheries and destructive fishing practices. While Day received support from several district officers, opponents to state regulation succeeded in delaying the matter for over 20 years, arguing that the implementation of regulative measures would entail too much expense. Others felt that fisheries were simply not important enough to deserve regulation, as they did not produce enough revenues. Thus, the Government of India repealed several moves by different provincial governments for the preservation of fisheries until the late 1880s. Eventually, the Indian Fisheries Act of 1897 was deliberately left malleable, and by 1904, only a number of provinces had made rules under it, which allows the conclusion that 'there was still considerable ambiguity in colonial attitudes to the question of regulating – and conserving – inland fisheries'.[99]

The first Indian debate around the introduction of a national river pollution law resembles the debates around forest and fisheries laws inasmuch as a group of officials voiced concerns over a particular resource, river water in this case, and sought to establish legislation to protect and preserve it. Arguably, their activism was neither driven by a financial motive nor by what we today would call ecological awareness. Instead, it was caused by anxieties over the potentially devastating effects of sewage-polluted river water on the health of riparian populations. Ultimately, these 'conservationists' were able to determine official policy on paper. However, they were not able to move the central government into passing a binding national law, as they did not represent majority opinion. The central government's refusal to pass an Indian Rivers Pollution Act can be explained in two ways. For one, the colonial ideology of difference, and, more concretely, the 'Indian paradigm', allowed colonial officials to ascribe much greater diluting and self-purifying capacities to 'tropical' Indian rivers, and thus to downplay potential dangers involved. Second, Indian rivers were indeed pristinely clean compared to their European counterparts, even though local pollution most certainly existed. Thus, the main incentive that led to forest, fisheries and wildlife legislation – overexploitation – was missing, since Indian rivers had not yet been turned into sewers. In the colonial pattern, conservation measures were always implemented *after* overexploitation had become clearly palpable, and thus India was to remain without a national water pollution law until after Independence.

The Kanpur sewage disposal committee's guidelines, issued on a background of highly divergent official opinion and lacking wider support, stood on shaky grounds from the very beginning. The remaining chapters will illustrate how, by neither passing national legislation nor enforcing the committee guidelines, the Indian government failed to support and promote non-polluting sewerage technologies during the decades that followed.

Notes

1 *British Medical Journal*, 8 February 1890, p. 311.
2 Ibid., p. 312.
3 Harrison 1980: 170–4. This allegation along racial lines mirrored similar allegations expressed along the lines of class in Britain, where elites claimed that the working classes with their 'dirty' habits needed to be 'civilised' (Mann 2004: 8).
4 Harrison 1994: 7–9.
5 Arnold 1993: 159.
6 Harrison 1994: 100–3.
7 Cuningham 1884: 130.
8 Harrison 1994: 102–4, 108; Kumar 1998: 174–5.
9 Huber 2006; Harrison 1996: 133–4.
10 Arnold 2000: 83.
11 Ogawa 2000: 675–6; Watts 2001: 344–51.
12 Harrison 1994: 35.
13 Isaacs 1998.
14 Chakrabarti 2012: 30–4, 142.
15 Harrison 1994: 115.
16 Klein 1980: 47.
17 Kumar 2013: 91–2; Chakrabarti 2014: 111–12.
18 Oldenburg 1988: xv–xvii. For a theoretical discussion on the 'colonial city' see King 1980. For Lucknow see Oldenburg 1988; for Delhi see Prashad 2001; Mann 2007; and Sharan 2014; for Allahabad see Harrison 1980; for Ahmedabad, see Gillion 1968.
19 Baber 1996: 200–12. Dalhousie moreover fundamentally reorganised the Indian postal system (ibid.).
20 Sharan 2011: 426.
21 Oldenburg 1988: xviii–xxi; Broich 2007: 349, 363–4.
22 Headrick 1988: 156–7; Dossal 1991: 95–148; Gandy 2008: 112–16.
23 Harrison 1994: 105.
24 Tinker 1968: 28–37, 43–8.
25 Harrison 1994: 172, 176–7.
26 Ibid.: 76–8; Oldenburg 1988: 96–144.
27 Colvin 1894: 516–20. See also Harrison (1994: 181–2, 193–4).
28 Colvin 1894: 516–20.
29 Freitag 2010a: 1.
30 Klein 1994: 500–5.
31 Arnold 2010: 254–6.
32 Eck 1993: 3–42, 211–20.
33 Fitzjames 1880: 9.
34 A *ghāṭ* is a flight of steps leading down to a river. It is often used as a platform for ritual bathing.
35 *Annual Report of the Sanitary Commissioner of the North-Western Provinces and Oudh [henceforth: ARSC NWP] for the year 1880*, p. 57.
36 *ARSC NWP for the year 1870*, pp. 79–80; Nevill 1909a: 264.
37 *ARSC NWP for the year 1886*, pp. 66–7.
38 Alley 2002: 158.
39 Committee for the Cleaning/Purification of the Ganga at Kashi.
40 On Raja Shivaprasad (1823–95), an influential scholar, Hindi writer, loyal government official and ardent promoter of the use of Hindi as administrative and educational language, see Stark 2012; Singh 2000: 355–6.
41 *ARSC NWP for the year 1886*, pp. 65–8.
42 'Proceedings of a public meeting held to concert measures to prevent the pollution of the River Ganges by the discharge of filth into it through the sewers', in *A Collection*

of Miscellaneous Papers Relating to the Benares Water-Works and Drainage Projects, p. 2, Uttar Pradesh State Archives [henceforth: UPSA], GoNWP, Municipal Dpt., File 3/63/1890, Box 532, 'Benares Water-Supply and Drainage Project'.

43 Ibid., pp. 2–3.

44 *ARSC NWP for the year 1886*, p. 65.

45 'Reply of His Excellency the Viceroy to the address of the Kasi Ganga Prasadini Sabha, December 1890', UPSA, GoNWP, Municipal Dpt., File 979A, Box 114, 'Application from the members of the Kasi Ganga Prasadhini Sabha, Benares, for His Honor's patronage'.

46 Colvin 1894: 516–17.

47 Hughes describes these depots as 'hemispherical masonry basins [...] with a pipe at the bottom connecting with the sewer, a water-pipe for flushing the filth down the drain, and a sweeper in charge' completing the arrangement. A.J. Hughes, 'Report on the drainage and water-supply of the city of Benares', s.d., UPSA, GoNWP, Municipal Dpt., Prgs February 1889, No. 11, File 3/63/1890, Box no. 532, 'Benares Water-Supply and Drainage Project'.

48 Ibid.

49 *The Englishman*, 16 January 1890, p. 4; *The Pioneer*, 22 January 1890, pp. 112–13.

50 W.H. Gregg, Sanitary Commissioner, Bengal, to Secy GoBeng, Municipal Dpt., 27.1.1890, National Archives of India [henceforth: NAI], GOI, Home Dpt., Sanitary Branch, Prgs February 1890, No. 78A. Until 1912, today's state of Bihar was part of the Bengal province, so at the time of writing Bengal bordered directly on the North-Western Provinces.

51 Ibid.

52 H.J.S. Cotton, Offg. Secy GoBeng, Municipal Dpt., to Secy GOI, 4.2.1890, NAI, GOI, Home Dpt., Sanitary Branch, Prgs February 1890, No. 78.

53 Richard Smeaton, Secy GoNWP, to Secy GOI, 27.5.1890, NAI, GOI, Home Dpt., Sanitary Branch, Prgs June 1890, No. 75. Auckland Colvin mistakenly speaks of 'town drainage', whereas the matter in question is sewage.

54 Hitherto, no detailed study on the development of water law or water pollution law in British India is available. In addition to the works of L.N. Mathur (1980) and Kelly D. Alley (2002), short outlines can be found with Cullet and Gupta (2009: 161–2) and Venkat (2011: 83–8). Venkat specifically focusses on water pollution.

55 Mathur 1980: 88; Alley 2002: 134–7.

56 *Indian Penal Code, 1860*, Lucknow: Eastern Book Company, 32nd edn, 2010.

57 Venkat 2011: 83–4.

58 The definition is put forward in section 268. See also Alley (2002: 133).

59 Venkat 2011: 83–4; Sengar 2007: 57–8; Abraham 1999: 44–6.

60 Alley 2002: 121–39; Venkat 2011: 84. In this context, Abraham (1999: 44) mentions one of the earliest recorded cases fought over river pollution in 1881, when the Madras High Court refused to convict one Vitti Chokkan for polluting the Varaga river with the explanation that the words 'public spring' and 'reservoir' in section 277 did not include a river.

61 Mathur 1980: 90.

62 Alley 2002: 135–6.

63 *ARSC NWP for the year 1879*, pp. 56–8. For similarly moralising accounts see *Calcutta Review*, Vol. X (1848), pp. 404–36 and *Calcutta Review*, Vol. XVI (1851), pp. 222–5.

64 Sanitary Commissioner, Madras, to Chief Secy GOI, 21.2.1877, in *Proceedings of the Sanitary Commissioner for Madras for the Year 1877*, p. 44.

65 British Library, Asia, Pacific and Africa Collections, India Office Records [henceforth: APAC, IOR], P/1003, GOI, Home Dpt., Sanitary Branch, Prgs March 1877 (B), Nos 2–3.

66 *Annual Report on the Administration of the Bengal Presidency for 1863–64*, pp. 88–91; APAC, IOR, P/146/71, GoBeng, Judicial Dpt., Prgs March 1864, No. 144.

67 *Annual Report of the Sanitary Commission for Bengal 1864–65*, pp. 62–3.
68 *The Indian Mirror*, 1 June 1864, p. 147; Supplement to *The Englishman*, 3 June 1864.
69 NAI, GOI, Home Dpt., Sanitary Branch, Prgs June 1890, Nos 75–6, Notes. The two publications W.R. Rice is referring to here are Edmund Parkes' *Manual of Practical Hygiene*, first published in 1864, and Baldwin Latham's *Sanitary Engineering*, first published in 1873. Parkes' (1819–76) *Manual* was initially conceived as a guide for medical officers in the army during his tenure as a professor of hygiene at the Army Medical School in Netley. It remained the standard work on military hygiene in Britain for much of the nineteenth century and continued to be edited after his death in 1876. As Parkes belonged to the pre-bacteriological era, he was chiefly concerned with the impact of environmental factors such as air, food and water on health. However, he did recognise the communicability of disease through intestinal discharges, even though he was not able to identify the nature of the transmissible material (Rosen 1976; Curtin 1989: 105). In India, Parkes was used as a reference by those asking for more specific, water-related measures against cholera, as he recommended the chemical examination and purification of drinking water (Harrison 1994: 102–3). Baldwin Latham (1836–1917) was one of the most renowned sanitary engineers of his days, and by 1868 had designed the sewerage, irrigation and waterworks for 15 English towns. His *Sanitary Engineering* on the construction of sewerage and house drainage was regarded as a classic (Insley 2004). Being published in 1873, it too belonged to the pre-bacteriological era.
70 NAI, GOI, Home Dpt., Sanitary Branch, Prgs June 1890, Nos 75–6, Notes.
71 Ibid.
72 Ibid.
73 Ibid. At the same time, the sanitary commissioner was well aware that 'to run into a more or less stagnant river millions of gallons of liquid sewage daily, through well-aligned drains, is not by any means the same things as allowing it to filter slowly through the soil on the banks' (ibid.).
74 C.J. Lyall, Offg. Secy GOI, to Secy GoNWP, 27.6.1890, NAI, GOI, Home Dpt., Sanitary Branch, Prgs June 1890, No. 76.
75 Bellwinkel-Schempp 1982: 134–8, 152.
76 G. Hutcheson, Sanitary Commissioner NWP, and A.J. Hughes, Supervising Engineer, Municipal Works NWP, 'Joint note on the sewerage of the city of Cawnpore, with special reference to the arrangements for passing the sewage through cantonments', n.d. [1892], pp. 2–5, NAI, GOI, Home Dpt., Sanitary Branch, Prgs April 1893, Nos 17–25.
77 Major-General E.H.H. Collen, Secy GOI, Military Dpt., to Secy GoNWP, 15.7.1892, NAI, GOI, Home Dpt., Sanitary Branch, Prgs April 1893, Nos 17–25.
78 Secy NWP, to Secy GOI, Home Dpt., 26.11.1892, NAI, GOI, Home Dpt., Sanitary Branch, Prgs April 1893, Nos 17–25.
79 NAI, GOI, Home Dpt., Sanitary Branch, Prgs April 1893, No. 17–25, Notes. As one official put it, 'everything must depend so much on the particular conditions of each place that I doubt if "general principles" laid down on the report of such a Committee would be of much use' (ibid.).
80 'Report of the proceedings of a committee appointed under the orders of the Government of India', 8.11.1893, APAC, IOR, P/4555, GOI, Home Dpt., Sanitary Branch, Prgs March 1894, No. 48.
81 Hutcheson's position will be discussed in detail in the next chapter.
82 Sharan 2011: 447. On the colonial ideology of difference see Metcalf 1994: 66–159.
83 This paradigm emerged as a scientific and literal construction from the late eighteenth century and was well established by the middle of the nineteenth century (Arnold 2006; Driver 2004; Arnold 2013: 5).
84 *Annual Report of the Sanitary Commissioner for Bengal [henceforth: ARSC Bengal] for the year 1889*, Appendix V, p. ciii. See also *ARSC Bengal for the year 1888*, Appendix V, p. 111.

85 See, for example: Guha 1990; Grove 1997; Reeves 1995; Rangarajan 2008; Rajan 2006: 56–61; Drayton 2000: 234–7.
86 Arnold 2013: 2.
87 Kumar *et al.* 2011b: 2; Tucker 2012: 5–8. Woodlands were primarily cut down to satisfy the needs of the imperial shipbuilding industry, and to give way for cash-crop plantations and revenue-producing agriculture. For a detailed study on early environmental anxieties and the development and circulation of conservationist ideas under the East India Company see Grove (1997) and Beattie (2011, ch. 4).
88 Guha 1990; Rajan 2006: 55, 57.
89 Rangarajan 1996: 55–137; Gadgil and Guha 1992: 113–80; Guha 1989.
90 Reeves 1995: 279–88.
91 Rangarajan 2008; Rangarajan 1998.
92 Beinart and Hughes 2007: 1–3.
93 Gadgil and Guha 1992: 113–45.
94 Grove 1997: 380–473.
95 Rangarajan 1996: 7.
96 Richard Drayton for example challenges Grove's sharp distinction between the histories of exploitation and conservation. He argues that the latter, while apparently being contradictory to the ethic of exploitation, was based on the same paternalist ideology (Drayton 2000: 234–5).
97 Reeves 1995: 279–88; Rangarajan 2008: 49–59. The protection of wildlife, for example, was increasingly viewed as a mark of civilised conduct (Rangarajan 2008: 56).
98 Guha 1990: 67–8.
99 Reeves 1995: 279–88 (quote p. 288).

2 River of disease

Considering the recent breakthroughs in bacteriological science, the Government of India remained strangely silent on the possible implications of sewage-polluted rivers for the spread of disease. In contrast, the government of the North-Western Provinces – temporarily given a free hand to frame its own policies by Governor-General Lansdowne in 1890 – put this question at the centre of debate. As the secretary to the North-Western Provinces remarked, the 'contamination question' was of great importance and had to be settled as soon as possible.[1] This, however, proved thoroughly complicated. The 'contamination question' stirred up a hornet's nest and generated a prolonged and heated debate, in which medical men and engineers fought over the nature of pathogens and an array of related questions: for example, whether and how they could be transmitted by rivers, whether a process of river self-purification existed, how it worked and how it acted on pathogens, and what method of water analysis may be used as a reliable tool to determine water quality.

So far, the historiography on public health and medicine in colonial India has not looked at the impact of germ theory on government policies on sewage disposal and related questions of water quality.[2] Scholarship on urban development, on the other hand, has overall focussed on the *extent* of water supply, drainage and sewerage in the colonial city, but not, as Awadhendra Sharan has rightly observed, examined debates and policies around water quality, narratives of pollution, purification technologies, and the transformations these all effected in a colonial context.[3] In this chapter, I contend that the still widely prevalent resistance to germ theory among colonial officials in British India during the early 1890s decisively shaped the North-Western Provinces' strategies on sewage disposal and river pollution. The provincial government's unwillingness to acknowledge the relevance of germ theory shaped its methods of inquiry into water quality and allowed it to deny the potential dangers ensuing from the discharge of untreated sewage into the Ganges. The 'Indian paradigm' served as another important justification of this practice. During the 1890s, the changing political context and the progress in bacteriological science did lead to a growing rapprochement of official opinion and methods of inquiry towards germ theory. But the basic orientation of river pollution policy – the choice of sewerage technologies along the criteria of simplicity and financial feasibility – did not change, as officials came to adapt the 'Indian paradigm' to the new disease theory.

The following analysis will concentrate on two main aspects: First, it will closely examine the argumentative strategies employed by different parties involved in the debate in their attempt to direct government action. During the early 1890s, two principal positions emerged, each dovetailing with distinct views on disease theory, and, connected to this, views on river self-purification and water analysis. Opponents to the Banaras project maintained that sewage contained pathological germs, and that its discharge into rivers would consequently lead to the spread of disease. They agreed that a process of river self-purification existed, but held that it had no meaningful effect on germs. Taking pathogens to be germs, they were moreover convinced that the only reliable method to assess water quality was bacteriological water analysis. In contrast, advocates of the Banaras project either disputed the existence of germs altogether, or stipulated that for several reasons, such as river self-purification or simply dilution, germs and other harmful elements ceased to be a threat once they entered the river. Critical attitudes towards germs went hand in hand with an assured confidence in chemical water analysis. While framing their argument, advocates of the Banaras project moreover construed a specific 'Indian paradigm', oriented along the physical character of Indian rivers and the alleged 'tropicality' of the Indian environment as a whole, which allowed them to rate sewage disposal into Indian rivers as far less problematic than sewage disposal into British rivers. After germ theory had gained definite ground, this argument was translated and adapted onto the new context. By the end of the nineteenth century, it was widely believed that the Indian environment, much more than the 'temperate' European one, effectively neutralised harmful germs in rivers.

The second important aspect to be examined is the role played by metropolitan and colonial experts. What kind of expertise did colonial officials resort to, directly and indirectly, to frame their arguments? What kind of experts ultimately shaped the decision-making process? As we shall see, the main point of reference for officials on both sides was metropolitan expert opinion, either by recourse to publications, or by direct involvement of experts present in India. At the same time, however, we can observe the emergence of colonial expertise on river pollution through sewage and other wastes in the person of Ernest Hanbury Hankin, the director of the North-Western Provinces' new bacteriological government laboratory established in Agra in 1892. The development towards a colonial expertise on river pollution, and much more so on sewage disposal, became more marked during the 1900s, and will be followed up in later chapters.

The most vocal opponent against Banaras's sewerage plans was the sanitary commissioner of the North-Western Provinces, G. Hutcheson. 'The cholera bacillus', Hutcheson wrote,

> exists in water, and can be preserved in soil for a considerable number of days: and the current of a rapid flowing river may carry it and other active organic particles hundreds of miles within a day or two. Sewage at its outfall

into a river is admittedly directly noxious and dangerous; and it can only be regarded at varying distances as less injurious not from actual dilution but from the chance of avoidance. Many of the epidemic diseases which prove fatal to human kind [*sic*] are of the zymotic type, and the little poison that may be disseminated in the vast volume of even a stream such as the Ganges may prove the seed of an extensive epidemic.[4]

Thus, sewage-polluted river water was likely to spread not only cholera, but other diseases too, such as 'tuberculosis, anthrax, erysipelas, tetanus, enteric fever, diphtheria, leprosy [and] relapsing fever', all of which were 'communicable and associated with a micro-organism of a constant and distinct species'.[5] The threat was especially grave in India, the sanitary commissioner warned, as rivers provided the main source of drinking water for the majority of riparian populations, and was generally drunk without filtration. The fact that the pathogens in question were germs, and not simple putrefying organic matter, carried two important implications. For one, any reference to river self-purification was clearly irrelevant, since

the theory of the oxidization of sewage in running water, and the self-purifying effect from other causes of rivers and streams, has reference more to the dead organic particles and not to the living germs of disease which undoubtedly find in sewage contaminated water the means of extended life and vitality.[6]

Secondly, chemical analysis provided no reliable means to determine the potability of water once it had come in contact with sewage: 'The chemical analysis of potable water is mere quackery, if it be intended to convey that a water presumably pure chemically is in all cases absolutely safe physiologically'.[7] Quoting C.H. Warden, professor of chemistry at Calcutta, Hutcheson questioned the relevance of measuring albuminoid ammonia, which chemists commonly took as an indicator for water pollution, in relation to disease. Warden himself emphasised the inability of chemical analysis to detect deadly micro-organisms. Thus, Hutcheson's position was evidently influenced by the recent breakthroughs of bacteriological science, but he also drew on discourses well established during the pre-bacteriological era. Quoting George Buchanan, medical officer to the British Local Government Board, Hutcheson adduced the concept of 'previous sewage contamination'. Buchanan, like Edward Frankland, believed that since it was impossible to measure 'dangerous organic pollution' with chemical water analysis, the origins of the water in question and the sources of pollution it was exposed to had to be taken into account when judging its wholesomeness. In an obvious reference to Edward Frankland, Hutcheson claimed that 'the water of any English river contaminated by sewage is unsafe for potable purposes from the point of contamination to the sea'. Generally, the sanitary commissioner found it hard to 'ignore the sanitary experience of all England, the recorded facts of the past half century, and the action taken by the English Parliament to remedy river pollution'. And while the Rivers Pollution Commissioners, Frankland and Buchanan had possessed only little knowledge about the exact nature of organic pollution, their position had now been vindicated by bacteriological science, which had furnished definite proofs as to the 'relationship between pathogenic bacteria and infectious disease'. Consequently, Hutcheson insisted that in addition to the chemical analysis of river water, bacteriological analyses were indispensable.[8] His memorandum ended with a passionate plea, which vividly brings out his deep commitment to bacteriology:

> Bacteriological science, in the person of Pasteur, has already extensively mitigated the ravages of disease in men and animals by his discovery of the processes of protective inoculation; in the person of Lister has relieved untold human agony by his discovery of the means to avoid suppuration and preserve the life of the wounded; and in the person of Koch has stayed death in the young by his discovery of the cause of the plague spot (phthisis) in the human breast and in searching out the secret whereby it may be effectively destroyed. These bridges reared by Pasteur, Lister and Koch over the *dark-flowing rivers of ignorance* are more palpable and enduring than the most solid masonry ever constructed by expert human hand.[9]

Hutcheson's views were strongly supported by the water analyst[10] to the Banaras municipality, W. Venis.[11] Venis, too, maintained that the dangerous elements in sewage-contaminated water were 'specific poisons of disease', and that water could therefore be considerably polluted even without containing large amounts of organic matter. Drawing extensively on Edward Frankland and the second Rivers Pollution Commission, he insisted that the harmful matter in sewage was neither neutralised by dilution nor by any process of river self-purification, that to detect it was quite beyond the scope of chemical water analysis, and that river water once polluted by sewage was not fit to be drunk anymore, especially not without filtration. Thus, both Venis and Hutcheson built their opinions on a combination between the most recent bacteriological knowledge and arguments from the pre-bacteriological era, in which the former testified to the validity of and reinforced the latter.[12]

In the meantime, Bengal's protest was reiterated by the Calcutta Health Society, a pressure group founded in Calcutta in 1884. Mainly comprised of Europeans and referring to a spirit of Christian duty and 'beneficent utilitarianism', the Society lobbied against the alleged neglect of public health under the municipality of Calcutta, which was dominated by Indians and liberal Englishmen.[13] In a letter to the Government of India, the body expressed its alarm about the Banaras sewerage project. If the principles laid down at Banaras were followed by other riparian towns, the Society admonished, the safety of municipal water supplies downstream and ultimately of Calcutta itself would be seriously compromised. Diseases like cholera and typhoid, it contended with reference to W.H. Gregg, were transmitted by germs present in excrement-polluted water, and these were neither destroyed by dilution nor detectable through chemical water analysis. Quoting the sanitary authority Albert H. Buck,[14] the Society concluded:

> [T]here is a liability of over-rating the amount of spontaneous purification to which a running stream is subject, and it is certain that we cannot decide with confidence as to when a stream once polluted becomes fit to drink; moreover as it is not possible by any practicable chemical treatment, or by any process of filtration to make a polluted water wholesome, it is safer not to use as a source of domestic supply a water which is known to have been seriously polluted.[15]

Thus, the opponents of the Banaras sewerage project were unanimous that sewage, before being discharged into the Ganges, had to be purified beforehand, either by filtration of the sewage through soil, its use for irrigation on a sewage farm, or a combination of both.[16]

The sanitary commissioner and others found their most articulate adversary in the originator of the Banaras project, sanitary engineer A.J. Hughes. A sewage farm, Hughes insisted, was quite beyond the financial scope of the Banaras municipality, and in any case could only be thought of once experience had been gained as to the quantity and quality of the sewage. As a temporary measure, the discharge of untreated sewage into the Ganges was therefore unavoidable. Hughes agreed that this practice was wrong 'in principle' and from a 'sentimental' point of view, but contended that no harm would come of it. '[T]he self-purifying properties of rivers', he asserted, 'are now strongly insisted on in every elementary work on sanitary science […] and there is no fact better established […] than the fact that organic matter is rapidly oxidized, diluted, absorbed and practically eliminated in rivers'.[17] The climatic and environmental conditions of India supported the different factors involved in river self-purification and intensified their effectiveness:

> In rolling over the broad sands of the Ganges under the blazing heat of a tropical sun, the oxidation and aeration of the water is carried on on a scale which never could have entered into the contemplation of English experts. The purification of sewage by deposit, by absorption, by plants, and by fish, and aquatic infusoria and deposit, are elements which it is difficult to give precise weight to; but that they are important factors in the case of a river 1,100 miles long will perhaps be admitted even by the Health Society.[18]

The key factor, according to Hughes, was dilution. British laws and the findings of the Rivers Pollution Commission could not simply be applied to India, as the water volume of British rivers was extraordinarily small in comparison to that of such rivers as the Ganges. Moreover, Hughes was well aware of the opposition Edward Frankland and like-minded experts were facing in Britain, and drew extensively on respective sources. Quoting William Santo Crimp[19] for instance, Hughes challenged the idea of fixed standards for effluent and water purity, holding that a state of 'pollution' was only reached when the amount of sewage poured into a river exceeded the self-purifying capacities of the latter. Taking the Banaras case, the engineer drew up a meticulous calculation involving the water volume of the Ganges, the projected amount of sewage to be discharged, the amount of the 'dangerous element' in the sewage and other factors, in order to prove that the latter two would be readily dissolved by dilution. Hughes' logic evidently built on the traditional assumption that the 'dangerous element' in sewage was some organic matter subject to putrefaction and decomposition. Countering arguments founded in bacteriology, he noted: '[t]he existence of specific pathogenic germs is by many regarded as altogether chimerical and fanciful, and the information about their life history and the circumstances under which they can exist and develop is vague and undefined.[20] Referring to Max von Pettenkofer's reckless auto-experiment, Hughes held up the 'distinguished biologist' who apparently 'swallowed several thousands of the cholera *comma bacillus* without ill effects'.[21] But even if the germ theory were true, the threat implied would still be negligible:

> [A] single germ might produce disastrous effects under favourable conditions in the recipient; but the chances under the conditions above described [the large water volume of the Ganges] are so remote and improbable that I shall be much surprised if a usually well-informed public body [the Calcutta Health Society] has really taken up the subject in [a] serious manner [...].[22]

The centrality of dilution in Hughes' perspective came to the fore very clearly during a conference held between the sanitary and engineering staffs of the North-Western Provinces and the Punjab, respectively, in November 1890. Following G. Hutcheson's suggestion, the conference was convened to discuss the potential pollution of Indian rivers by municipal sewage. As it turned out, the Punjab government itself was in the middle of devising a new drainage scheme for Delhi, one, however, in which only surface drainage was to be discharged into the river Yamuna, while street sweepings and solid excreta were still to be trenched. Naturally, the drainage would contain a certain amount of organic waste.[23] During the meeting, Hughes took exception to this proposition, arguing that the drainage contained enough faecal matter to turn the river water into a danger for the riparian villages below Delhi (including those situated in the North-Western Provinces) during the hot weather, when the water volume of the river was very low. As the conference notes suggest, the meeting came to solely revolve around this concern, rather than to engage with river pollution by municipal sewage in general terms. Regarding this question, too, the meeting yielded no direct results, as opinions about the pollution question and alternative means of sewage disposal remained irreconcilable. However, Hughes and his colleagues managed to move Punjab into considering the disposal of the Delhi drainage on land.[24] Their effort seems to have been successful, since in 1893, the Punjab government intimated the North-Western Provinces of its decision to dispose of the Delhi drainage on land, to be used for broad irrigation.[25]

For all the debate around public health and germs, Hughes' position was driven by a strong financial motive as well. Assuming that there was no pressing need for sewage purification, Hughes thought it unwise to burden municipal funds, already strained by the expenditure for basic waterworks and sewerage systems, with additional costs for sewage treatment works. The Banaras municipality, Hughes pointed out, would neither be able nor ready to raise 55,000 to 76,000 rupees annually by taxation for the latter, in addition to the 400,000 that had to be raised for the sewerage system itself. To put such pressures on municipalities and tax payers may have the adverse effect of causing them to reject any measure of sanitary reform. Hughes' concluding statement in this context is worth quoting, as it succinctly conveys the spirit of colonial policies on urban sanitation in British India:

> We must be content to obtain the most important essentials of sanitation in the shape of a pure and abundant supply of water, well distributed in the city, and the best and most elastic system of sewers we can afford to pay for: further than this, it is not possible to go.[26]

Hughes received influential backing from J. Richardson, Inspector-General of Civil Hospitals of the North-Western Provinces.[27] Richardson maintained that Gregg had furnished no proof whatsoever that the sewage presently poured into the Ganges at Banaras and elsewhere through the existing drains left behind any measurable pollution, nor that this was causing sickness in the population. In a failed effort for a joint memorandum with G. Hutcheson that ended in a heated exchange between the two, Richardson moreover asserted that the main source of drinking water of the riparian population was not the Ganges, but wells. As per Hutcheson's warning regarding epidemic disease he stated:

> The 'cholera bacillus' is dragged into the discussion, for the sake, it is presumed, of the moral effect which its name may produce. But apart from the circumstance that scientists are still not altogether agreed regarding the existence of a specific 'cholera bacillus' the suggestion that cholera may be spread by the discharge of sewage into the Ganges at Benares is particularly infelicitous. No one knows better than Dr. Hutcheson that the general course of cholera in these provinces is from east to west, or up-stream.[28]

Alike Hughes, Richardson agreed that the discharge of untreated sewage into the Ganges was 'objectionable', and also 'unscientific', in principle. But in his view, Hutcheson and others exaggerated the possible harm that could result from it, especially if it was done only temporarily. His reply to Hutcheson's extensive memorandum was short and crisp. Without responding to the various points discussed, he simply asked his colleague to suggest an alternative scheme of sewage disposal, which was 'feasible and capable of being carried out with the available funds'.[29]

This heated debate about the impact of sewage on river water quality reflects the controversial status germ theory occupied in British India during the early

1890s. Official policy, prominently espoused by D.D. Cunningham at Calcutta, still considerably held on to localist explanations of cholera. While Cunningham had come to accept the existence of germs, he fundamentally questioned the causal relation between the disease and the germ Koch had isolated. Deeply influenced by Max von Pettenkofer's work, Cunningham maintained that were there were several different germs related to cholera and that the characteristics of each type were dependent on local conditions.[30] In Britain, too, the discoveries of Koch and others had been received with a great deal of scepticism. At the forefront of British opinion stood the bacteriologist Edward Klein, who was strongly supported by the British government and a majority of medical health officers. In 1884, Klein had visited Calcutta together with Heneage Gibbes and Alfred Lingard as member of the British cholera commission to investigate Koch's claims. In his report submitted in 1885, Klein, like D.D. Cunningham, closely followed the Pettenkofer model of disease aetiology in denying any direct causative relation between germ and disease. Instead, he insisted that the real cause of cholera was a chemical virus produced by a disease agent bred in a suitable soil.[31] Thus, both in Britain and British India, many aspects of germ theory remained open to debate, and, as the London water controversy of 1892/93 shows, germ theory did not provide clear guiding principles for water quality questions in Britain itself. On these lines, Hughes stated that 'there is at least as much probability that the strongly alkaline sewage would kill germs as that they could develop in sewage'.[32] As a proof of the insecurities surrounding germ theory, Hughes appended an extract from a lecture held by Robert Koch in front of the International Medical Congress in Berlin in 1890, and commented that '[i]t appears to show how very little we know, and how little reason there is to speak of pathogenic germs in the confident and positive way so often used by persons who are not experts'.[33] The North-Western Provinces government, confronted with two irreconcilable factions, decided relatively early to rely on the authority of its inspector-general of civil hospitals. This was not least because, as the secretary to government made it clear, his opinions coincided with government's own.[34] Richardson belonged to the old generation of IMS officers, having entered the service in 1859. In 1866, he joined the North-Western Provinces' sanitary department as deputy sanitary commissioner, later becoming sanitary commissioner, and finally inspector-general of civil hospitals in 1890. According to his obituary in the *British Medical Journal*, Richardson was 'very popular and also a strong head of the medical department of that province'. In 1887, he represented India at the international sanitary conference in Venice, and in 1897 officiated as a member of the Indian Famine Commission.[35]

Following Richardson's advice, the North-Western Provinces sought to determine the existence or absence of a public health threat by investigating two principal points. First, Richardson had contested the assumption that river water provided the main source of drinking water for the riparian population along the Ganges, and second, the assumption that consuming Ganges water mixed with sewage could be a cause for sickness.[36] Therefore, government directed the commissioners of the Agra, Allahabad and Banaras divisions to conduct surveys to find out to what extent river water was used for drinking in and around the cities of Kanpur, Allahabad, Banaras, Mirzapur and Agra. Moreover, they were directed to collect information as to whether river water was considered to cause sickness generally or at certain times of the year. This was to be done through the municipal boards, and, where villages were concerned, through the *kanungos* during their regular inspection tours.[37] The survey's results were taken by Richardson as a confirmation of his views. According to the inspector-general, they proved that the majority of people used wells for their main source of drinking water; in Banaras 'only' one fifth of the population (which, one may note, were still 40,000 people, taking a total population of 200,000) depended on river water, in Mirzapur one fourth, and in Kanpur one person in 25. Moreover, they showed that the people who used river water did not fall victim to epidemic disease, or were generally more often affected health-wise, that those who depended on wells.[38] On the basis of Richardson's judgement, the North-Western Provinces government considered the matter settled.[39] A closer look at the forms returned by the commissioners, however, puts Richardson's evaluation to test. Most importantly, the populations of Banaras and Mirzapur regarded Ganges water as less wholesome at and near points where drains discharged into the river. In Mirzapur, the waste water from houses and lac factories was directly made responsible for the increased incidences of cholera, bowel complaints, fever and other diseases during seasons other than the rainy season. And the results for Allahabad, not referred to by Richardson, showed that every second inhabitant used Ganges water, the other half citing distance from the river as the reason why they resorted to wells. In fact, one Allahabad official ascribed the widespread use of Ganges water and its perceived wholesomeness to the fact that no wastewater of any kind fell into the Ganges for many miles above the city. In the other towns, too, the distance from the river seems to have been a major criterion according to the forms, since the majority of those who lived close enough drew their drinking water from it.[40]

While the survey was still in progress, A.J. Hughes with the support of government tried to strengthen his position by seeking the advice of metropolitan experts. One of these was the eminent agricultural chemist John Augustus Voelcker, consulting chemist to the Royal Agricultural Society of England. In 1889, Voelcker had come to India by invitation from the Government of India to report on the state of Indian agriculture.[41] At an informal meeting with Hughes and the secretary of the North-Western Provinces, Voelcker stated that the temporary discharge of untreated sewage, even though objectionable, was not likely to cause any harm.[42] Additionally, Hughes set out to summon some of the most renowned British experts in the fields of sewerage and water quality questions to fight his case. In a letter to Baldwin Latham,[43] Hughes expressed his desire to procure reports on the Banaras sewerage question by Latham himself, as well as by Charles Meymott Tidy, William Corfield and James Wanklyn, and inquired into the costs at which such reports could be obtained.[44] In his reply, Latham suggested that the Banaras case be assessed by William Corfield, Charles Meymott Tidy and William Odling, 'who has combated Dr. Frankland's views as to the oxidation theory'.[45] While no reports were ultimately arranged from the latter three, Latham himself visited India a couple of months later on an invitation by the Bombay and Calcutta municipalities to advise on their drainage problems.[46] The North-Western Provinces used the opportunity and hired Latham to assess the technicalities of the Banaras and Kanpur waterworks and sewerage projects, and to report on the sewage disposal question. Reporting on Banaras in December 1890, Latham wrote:

> I am quite aware that the objections to running crude sewage into a quick-flowing river of large size are more sentimental than real, and in no case within my knowledge has any bad result occurred to the health of those living on the banks of a slightly polluted stream, or has the water, taken from a river receiving sewage after a fairly long flow, ever been proved to be the cause of disease. On the other hand, we have the actual experience that the flow down a river will destroy germs of disease. [...] Compared with the flood volume of the Ganges, the whole of the impurities that can be put into it are infinitesimal, and the provision of Nature in the living waters of an ever-flowing stream are so great as to destroy, in the course of a few miles' run, any dangerous impurities passed into it.[47]

Latham's confidence as to the discharge of untreated sewage into the Ganges appears remarkable when comparing his statement to a passage in his guide-book *Sanitary Engineering*, first published in 1873:

> [A]s it is advisable that the sewage in no case should be allowed to intermix with the pure natural water of the country, so as to lead to its pollution, provision must be made for either purifying the sewage before passing it into the fresh-water streams, or, as in the case of sea-coast towns, to lead it to such a point as not to become the cause of offense. In inland towns it will be found that there are chemical or mechanical systems which will greatly palliate the evils of pollution by precipitating or deodorizing the sewage, but the nuisance arising from sewage pollution may not always be removed in this way; consequently such works are generally supplemented by intermittent filtration, or irrigation works. The plan that has hitherto proved most successful in purifying the sewage of an inland town is that of utilizing it in its fresh state on properly prepared land.[48]

Evidently, Latham had fundamentally reversed his position in the late 1880s and by 1890 completely adhered to the theory of river self-purification, believing moreover that germs got destroyed while flowing downstream. While he considered the disposal of untreated sewage into rivers used for domestic water supplies as wrong in principle,[49] the volume and length of the Ganges in Latham's view guaranteed sufficient self-purification. Thus, the Ganges's physical character, arguably representative of that of many Indian rivers, made it appear as a suitable receptacle for untreated sewage.

The third step the North-Western Provinces undertook to assess the potential threat to public health posed by the Banaras project was the analysis of river water as directed by the Indian government. In the eyes of critics, this measure was essentially flawed from the outset. For one, as even the sanitary commissioner of India in suggesting the analyses had admitted, the present discharge of sewage into the Ganges at Banaras could not be compared with the organised discharge of the sewage of 200,000 people. Speaking from his experience on the spot, the water analyst to the Banaras municipality insisted that the amount of sewage finding its way into the Ganges through the old drains was very small, except during the rainy season. The old drains were faulty in construction and fall, mostly pervious, and there was no arrangement for flushing them, so that for eight months of the year the sewers were blocked with stagnated sewage and deposits.[50] The same was held true for the surface drains in the cities of Kanpur, Mirzapur and Lucknow.[51] Second, the usefulness of chemical analysis itself was doubted. As already discussed, Hutcheson, Venis and others believed that chemical analysis was ill-suited to detect the harmful components in sewage-polluted water, namely, germs of disease.[52] For the time being, their criticism went unheard, and government in 1892 initiated a series of chemical water analyses at Kanpur and Lucknow, executed by D.W. Aikman, resident engineer at the Lucknow waterworks.

In March 1892, D.W. Aikman submitted his first report. In both cities, water samples had been taken above and at several points below the principal sewers and drains discharging into the rivers. By sending a float downstream, the contribution of each outfall was assessed. As both Kanpur and Lucknow disposed of their excreta by hand-removal and trenching, the effluent in question was in fact sullage, even if it can be assumed that it contained a certain amount of excreta. This mostly consisted of municipal sullage, but also included factory effluents (e.g. from distilleries, paper mills and, especially in Kanpur, tanneries). The samples were sent to the water analyst at Banaras where they were tested for hardness, solids, chlorine, ammonia, nitrates and heavy metals. The Lucknow samples showed a steep incline in albuminoid ammonia as the Gomti passed the city. Calculations indicated that this was not due to the sullage, but caused by the bathing of men and animals, the washing of clothes, and percolations from sullage-sodden soils. In fact it was noted that the bulk of the city sullage was sinking into the soil and only a negligible amount found its way into the river at all. Similarly, the increase in free ammonia below the city of Kanpur was ascribed to the extensive cultivation of melons along the Ganges, which were extensively manured and irrigated.[53] Thus, chemical analysis revealed that the pollution caused by city sullage in both cases was negligible. Moreover, as D.W. Aikman emphasised, the experiments proved that the Gomti and the Ganges rapidly purified themselves from this pollution. After five miles of flow in the case of the Gomti, and after only one mile of flow in the case of the Ganges, the river water had reverted to its original purity and no remnants of pollution could be detected. To Aikman, this was proof of the extraordinary self-purifying abilities of Indian rivers. In Britain, Aikman wrote in his report, rivers were small and almost every riparian town and village was equipped with a water supply and sewerage system. They therefore quickly reached what he called their 'sewage saturation point', i.e. the point up to which a river based on its volume was able to purify itself naturally from pollutants. In India, where rivers were large and sanitary engineering widely absent, this point would only be reached in a remote future, when all the larger riparian towns discharged their sewage into the rivers. But long before this, Aikman put forward optimistically,

> Indian sanitation will have advanced, and our big towns, knowing the value of sewage as a fertilizer, will pump it to the nearest barren ground, and the revenue derived from their sewage farms, if it does not more than pay, or pay the expenses of pumping, will at least reduce those expenses to a mere nominal figure compared to the great advantages of a perfected sewerage system.[54]

Other than Hughes and Richardson before him, Aikman acknowledged the existence of pathogenic germs and the possible danger emanating from their presence in sewage-contaminated water. Yet, drawing on a study carried out in 1890 by the Italian scientist Alessandro Serafini at Max von Pettenkofer's Institute for Hygiene in Munich, Aikman contended that rivers not only rapidly purified themselves of organic matter, but also of harmful bacteria. Quoting Serafini, he wrote:

> The self-purification which is observed in rivers does not take place by means of oxidation in the mere volume of water; but it is due to a complex variety of factors. To sedimentation, to dilution, to the mechanical action of suspended substances which are deposited on the bottom, to movements of the waters, to lower temperature in the rivers than in the sewers, to superficial filtration on the bed of the river, and perhaps also to a certain action of the water per se, is due the rapid diminution of the bacteria which are contributed to rivers by sewerage, and to sedimentation, to dilution, and to slow and continual processes of oxidation in the bed of the river, is due the self-purification from organic substances and from the intermediate products of their decomposition in suspension and in solution in water.[55]

While Aikman believed that the volume of Indian rivers guaranteed a more rapid self-purification from organic matter, he cautioned that the hot climate of the country may increase the life span of bacteria, enabling them to live on even after the organic impurities, which had 'nourished them and gave them birth', had disappeared. He therefore suggested that, unless bacteriological analysis showed that pathogens died *before* the complete disintegration of all organic matter, it would not be safe to rely on chemical analysis alone, and bacteriological analyses should always be included.[56] The results and conclusions arrived at through the chemical analyses of Ganges and Gomti water in 1892 were overall confirmed by further samples taken in 1893 and 1894.[57] Additionally, samples were taken from the Yamuna below Delhi, in order to assess the impact of Delhi sullage on the Agra water supply. As in the case of the Ganges, the results gave no reason for concern.[58]

From 1894 onwards, water samples were also tested bacteriologically. Acting on the second recommendation put forward in the Government of India's resolution of 27 July 1888 – the promotion of facilities for scientific research and experiment – the North-Western Provinces in 1892 created a new government laboratory at Agra and appointed Ernest Hanbury Hankin as chemical and bacteriological examiner. This was a significant departure from earlier policies, when chemical examiners used to be drawn from the provincial staffs of medical officers. With Hankin, the North-Western Provinces also called for an expert bacteriologist for the first time.[59] Hankin, born in 1865 as the son of a clergyman, had received his medical education at London University College and St John's College in Cambridge, and after studying at St Bartholomew's Hospital for some time proceeded to work in Robert Koch's laboratory in Berlin and the Pasteur Institute in Paris. Before leaving for India, Hankin's research was mainly concerned with the germicidal action of alexins, i.e. proteins of blood-serum and cells, on which he published several papers in renowned journals such as the *Proceedings of the Royal Society* and the *Centralblatt für Bakteriologie*.[60] On his arrival in India, Hankin quickly became involved in the movement for a Pasteur Institute in India. In 1893, he joined the central Pasteur Institute committee and helped organise a meeting at Lahore, where Pasteur-enthusiasts pressed for the creation of an anti-rabies Pasteur Institute in India (which was eventually established in 1900 at Kasauli).[61]

After an initial delay, due to a lack of laboratory accommodation and suitable reagents,[62] Hankin embarked upon a detailed investigation into the microbial contents of the Ganges, the Yamuna and the Gomti in 1894 and 1895, the results of which he recorded in his annual reports to the government of the North-Western Provinces.[63] In 1896, he published an extended report on his research in a paper titled 'L'action bactéricide des eaux de la Jumna et du Ganges sur le microbe du choléra' in the *Annales de l'Institut Pasteur*.[64] At the outset of his experiments, Hankin had been surprised to find that both rivers contained a far smaller number of microbes than European rivers of comparable size, despite being subjected to various sources of pollution, such as mass bathing during religious festivals, the washing of clothes and cattle, and the disposal of large numbers of un- or insufficiently cremated cholera corpses. With respect to the latter, Hankin thought it remarkable that cholera epidemics did not spread downstream following the course of the Ganges, but from Bengal upwards. This, he noted, was one main reason for the long-time reluctance among medical men in British India to accept that cholera was a waterborne disease.[65] There existed a number of rather obvious reasons for the rivers' relative bacteriological purity: The near absence of sewerage systems in riparian cities, the comparatively small amount of industrial effluents, and the low density of settlement, which allowed the rivers to recuperate during long, pollution-free stretches. Like many before him, Hankin also believed that the rivers' self-purificatory powers were enhanced by the Indian climate:

Le pouvoir auto-purificateur qui dépend de l'action de l'air et de la lumière doit sans doute être beaucoup plus actif dans l'Inde qu'en Europe. Les larges rivières de provinces du N.W. courent en couches minces et sinueuses au milieu de bancs de sable, et sont dans de bonnes conditions pour éprouver l'action de la lumière et de l'oxygène, qu'aide l'action d'une temperature plus élevée qu'en Europe.[66]

However, Hankin found that the Ganges and the Yamuna possessed one more self-purifying characteristic, which made them unique not only against European, but also other Indian rivers: A mysterious germicidal property, an antiseptic which destroyed cholera and typhoid microbes in less than three hours. What exactly this germicidal agent was, Hankin was unable to say. The only thing he was sure of was that it did not enter the rivers from the outside:

La propriété antiseptique [du Jumna] ne lui vient donc ni du l'eau de fonte des neiges ni de l'eau de surface, elle est due à une substance inconnue formée dans le fleuve, ou recueillie par lui *in situ*. La même substance parait être présente dans le Gange. [...] Je ne puis rien ajouter sur la nature et l'origine de cette substance: je ne sais qu'une chose, c'est qu'elle est volatile.[67]

Notably, this germicidal power got greatly reduced or even destroyed when the river water was boiled at high temperature.

In a further experiment, Hankin tried to assess whether the Yamuna's germicidal powers were negatively affected by the pollution it received in and around Agra, especially by corpses of cholera victims floating in the river. The experiment was executed during the particularly dry winter months 1895/96, when the river's water mark had been rendered exceptionally low, and pollution was accordingly concentrated. Venturing out into the river on a boat, Hankin drew samples from above and below the city, and right next to floating corpses – an uneasy task, as these were quickly being consumed by large turtles. Evaluating his samples, Hankin found that the germicidal powers of the Yamuna were not affected, neither by the general pollution added by the city, nor by the corpses, and that after a maximum of three hours, all the cholera microbes had disappeared even in those samples taken near a freshly disposed cholera corpse. In what presents a remarkable statement, in view of the long-lived European outrage at the disposal of corpses into Indian rivers, Hankin concluded that even though '[t]out ce qui accompagne cette coutume est répulsive pour nos goûts européens [...] les résultats [...] montrent qu'il n'y a pas d'objection pratique à faire à cette coutume'.[68] So convinced was Hankin of the germicidal powers of the two rivers that he recommended the authorities to encourage pilgrims at holy Hindu sites to drink from them, and to discourage the use of wells.

Based on the results of his analyses, Hankin drew important conclusions with regard to sewage disposal into rivers. In his 1894 report to government, Hankin noted: 'If it is admissible to throw sewage into European rivers, *à fortiori* it is admissible to dispose of sewage in this way in these Provinces.' Nevertheless, he was hesitant to apply his results to any given river. While the Ganges and the Yamuna purified themselves rapidly due to the presence of various conducive factors, excessive disposal of sewage into a small river such as the Gomti at Lucknow might be objectionable. Another decisive factor was the density of populations below sewage outfalls. Moreover, it was possible that the Ganges and the Yamuna might lose their germicidal powers if they became highly polluted. Definite answers to questions like this, Hankin concluded, could only be given on the basis of a long series of experiments carried out on the respective rivers, and could not be deduced from experiments carried out on European rivers.[69]

The practical implications of Hankin's reports were clear enough: Generally, the Ganges and the Yamuna possessed extraordinary self-purificatory and, more specifically, germicidal powers, which rendered the discharge of sewage into them largely unproblematic, especially when the land near sewage outfalls was thinly populated. Where rivers were small and human settlements existed near sewage outfalls, a more cautious policy was advisable. As we shall see, the North-Western Provinces' official policy on sewage disposal into rivers largely followed Hankin's model during the years to come.

The debate around river pollution through sewage in the North-Western Provinces took place at a crucial junction in the history of official attitudes towards disease theory in British India. The initial phase of the debate, spanning approximately two years, reflects the pattern of resistance colonial officials had long displayed to contagionist theories in general and the germ theory in particular. This pattern found most prominent expression in J. Richardson, whose opinions carried considerable weight with his superiors. Joining the IMS in 1859, Richardson belonged to the old generation of IMS officers, the great majority of whom strictly clung to localist, miasmatic explanations of disease even after bacteriology progressed by leaps during the 1880s. While Richardson dismissed germ theory to the extent that he was unwilling to discuss any issue at stake in detail, his closest ally A.J. Hughes showed himself open to calculate with a hypothetical cholera germ, but his contributions make it sufficiently clear that he, too, strongly doubted its existence. Even though there existed an overall consensus about the validity of germ theory in contemporary European medical circles, Richardson and Hughes did not stand completely outside the mainstream of medical opinion. Their critical reception of germ theory tallied not only with the official doctrine of the Government of India, but also with that of the most influential bacteriologist and the majority of medical men in Britain. At the beginning of the 1890s, numerous details about germs and their behaviour under different conditions – how sewage and river water affected their survival and multiplication, or what combination of germs, predisposition and environment was necessary to cause an outbreak of disease, to repeat just a few – were still openly debated, and bacteriology was unable to provide clear guiding principles for policy-makers who had to decide on questions of water quality.[70] The river pollution debate in British India was therefore no less controversially fought than the debates taking place in the metropole itself.

While the Indian debate was constitutively shaped by the use of arguments employed in metropolitan debates – and thus reflects the latter in many ways – it acquired distinct forms due to the colonial context. Thus, advocates of the Banaras scheme construed an 'Indian paradigm' with regard to sewage disposal into rivers. The physical character of Indian rivers, especially their water volume, and the alleged 'tropicality' of the Indian environment as a whole were taken to significantly enhance the rivers' self-purifying powers and to render the disposal of untreated sewage into them unproblematic. The argument of river self-purification thus gained particular strength in the Indian context.

The 'Indian paradigm' built on certain long-lived tendencies of European observers and colonial officials in their perception of the Indian environment, and its relation to disease in particular. As David Arnold and others have shown, natural history and the natural sciences from the late eighteenth century began to forge a concept of the 'tropics' in distinction to 'moderate' or 'temperate' zones. While the 'temperate' zones represented everything modest, civilised and cultivated, the 'tropics' were marked by a dichotomy themselves: On the one hand, they appeared as almost dream-like places of Edenic plenitude and fecundity, home to enchanting and curious flora and fauna, and an abundance of minerals, precious stones and natural resources. On the other hand, the intensity of heat and humidity, the presence of diseases unknown in kind or intensity in Europe, insect pests and other phenomena testified to abundance and excess in a more troubling form, explaining the contemporaneous designation of the 'tropics' as the 'torrid zones'.[71] These negative connotations gained in strength the more Europeans penetrated and settled in the 'tropics', and with the arrival of 'Asiatic cholera' in Europe, the 'tropics' in general and India in particular came to be viewed as unclean and unhealthy, while disease was firmly linked with cultural and social backwardness.[72] The statements of A.J. Hughes and other officials about the purifying or purification-enhancing effects of the 'blazing heat of a tropical sun', the monsoon rains, and the extraordinary length and water volume of the Ganges illustrate how their judgement was decisively influenced by this concept of 'tropicality', which here was applied in a positive sense to sewage and water quality questions. A second pre-existing conception with which the 'Indian paradigm' resonates is the idea of the Indian environment being epidemiologically unique, due to the special meteorological conditions prevailing on the subcontinent. Based on this assumption, J.L. Bryden and other medical men had concluded that disease followed different laws in India than in Europe and therefore called for different public health policies.[73] Even though none of the statements makes direct reference to it, it may be theorised that this intellectual tradition, too, was still active in the background and supported the idea of sewage disposal into Indian rivers being less harmful than in Europe.

The 'Indian paradigm' was translatable to the new context of germ theory, once the latter became increasingly accepted in British India in the course of the 1890s. D.W. Aikman's statements show that this venture was initially accompanied by insecurities. While Aikman believed that pathogenic germs were destroyed by river self-purification much like any other organic substances, he was wary of the live-enhancing effect the hot climate of the country may have on them. As a look at Pratik Chakrabarti's recent research on the development of bacteriology in British India suggests, Aikman did not stand alone with his apprehensions. Germs in the 'tropics' acquired new meanings from those they carried in Europe. While the new science was finally received with great enthusiasm in British India towards the close of the nineteenth century, raising optimism as to the fight against disease, it also triggered anxieties. Questions emerged, such as whether hot climates influenced the behaviour of germs, whether 'tropical' environments caused germs to ferment or to become especially virulent, and whether vaccines could be produced and preserved in this environment. In fact, germs served to reinforce many of the negative assumptions about the tropics as pathological places, both physically and morally.[74]

The 'Indian paradigm' was also applied by E.H. Hankin, when he wrote about the self-purification enhancing characteristics of the Indian environment, and the special and mysterious germicidal property of the Ganges and the Yamuna, which according to him rendered them unique against European rivers. By the turn of the century, the discourse established by Hankin and others had struck roots, as it became widely believed that the 'tropical' Indian environment had a particular neutralising effect on germs. Expressing his opinion on the proposed discharge of sewage into the Ganges at Kanpur, the new sanitary commissioner of the North-Western Provinces wrote in 1898:

> I am very diffident in advancing any positive opinion on the general question of the risk of introducing crude sewage into a river [...]. Still I venture to believe that quite recently investigations and inquiry have led to a much greater value being attached than formerly to the germicidal powers of sunlight and free oxygenation; and it is certain that the conclusions of the River Pollution Commission in England have only a limited application to the circumstances of large Indian rivers, teeming with active organic life which feeds on impurities and flowing under an Eastern sun.[75]

As we shall see further down, the 'Indian paradigm' perpetuated official perspectives beyond the North-Western Provinces and the Government of India, and was employed by Bengal officials during the septic tank controversy in Calcutta as well.

The second issue calling for a closer analysis, addressed in the introduction to this chapter, is the role of metropolitan and colonial expertise in determining early colonial river pollution policies. The picture will become much clearer and will be discussed in detail after expert involvement has been further tracked into the 1900s; nevertheless it is useful to delineate some first contours here.

Questions about the effect of sewage on river water quality called for the expertise of men versed in chemistry and bacteriology, and the application of these sciences to water. Chemical water analysis in British India was routinely carried out by medical officers employed as chemical analysts, and was therefore already well-established.[76] Bacteriology started to gain definite ground in the colony from the early 1890s, but during the initial years of that decade, local expert bacteriologists were still very rare. D.D. Cunningham's laboratory at Calcutta remained unique for many years, until the opening of a bacteriological department at the Poona College of Science under the direction of the 'imperial bacteriologist' Alfred Lingard in 1891.[77] Scientific research in British India generally lagged behind contemporary standards of modern science, its objectives having been defined along administrational needs and economic utility since Company rule. India's isolation from modern science and the political urgency to harness science more effectively to the state was first recognised by Governor-General Lord Elgin (1894–99). In 1898, amidst growing popular unrest over how the state handled the Bombay plague epidemic, Elgin noted that in many fields, the Indian government possessed no competent advisors of its own. Instead, Elgin complained, scientific investigations were conducted by officials 'who are neither by experience nor knowledge competent to offer a decided opinion as to the best course to be pursued'. The result, all too often, was 'misdirected action' and fruitless endeavours.[78] He therefore called for a body of leading scientists to guide and control scientific research carried out in India. In 1899, the Indian Advisory Committee (I.A.C.) was created as a standing body within the Royal Society in Britain. The I.A.C. remained dormant for several years however, and in 1902 Governor-General Lord Curzon launched an Indian counterpart, the Board of Scientific Advice (B.S.A.), which was to unite the heads of all the ten Indian scientific services and thus ensure the coordination and efficiency of scientific research.[79]

In the absence of experts on the spot, metropolitan expertise on sewage and water quality during the initial years of the debate played a fundamental role in informing official opinion. Having no earlier Indian debates as a reference point and no own experts at hand, colonial officials turned to Britain for orientation. And as metropolitan debates themselves were highly controversial, it is little surprising that the Indian debate took on equally controversial forms. Advocates and opponents to the Banaras sewerage project both were able to draw on a range of contradictory expert opinions, choosing them selectively to give weight to their argument. At the same time, attempts were made to directly incorporate metropolitan expertise in the policy-making process, by consulting John Augustus Voelcker and Baldwin Latham. However, this happened more by 'accident' and was far from presenting a systematic inquiry. Voelcker was quite obviously addressed simply because he happened to be in India at the time, and while he was an agricultural chemist of great renown, he had not previously been involved with questions of river pollution through sewage to any significant extent.[80] In this regard, Latham was a more well-founded choice, even though it is not sure whether the North-Western Provinces would have called for him had he not been invited to India by the municipalities of Bombay and Calcutta. A principal reason why the North-Western Provinces did not involve metropolitan experts on a more extended scale was their reluctance to pay for them. Instead – and this is a point that will be discussed in greater detail in the next chapter – the provincial government put the financial responsibility on the Banaras Municipal Board, which in turn refused to accept this burden. Therefore, while *indirect* metropolitan expert involvement was strong, the actual scientific experiments and surveys to assess the impact of untreated sewage on rivers were initially carried out by the North-Western Provinces' own sanitary and engineering staff. Following the appointment of E.H. Hankin, inquiries into sewage disposal and river pollution were put on a much more sound scientific footing. Not only was Hankin one of the few highly trained bacteriologists available in India at the time, he was also the first among them to develop special expertise on river pollution.

The early river pollution debate's strong metropolitan focus is noteworthy. With very few exceptions, colonial officials took recourse to British administrative and scientific discourses on sewage disposal, river pollution and related questions of water quality. Experts actively consulted were equally British (Voelcker was of German descent, but lived in Britain from an early age).[81] In contrast, international discourses on these issues received little attention. This is most clearly illustrated by the fact that a contemporary controversy revolving around Munich's sewerage system was virtually ignored. In the late 1880s, Freising, Landshut and other riparian towns launched a protest against the projected discharge of Munich's excreta into the Isar. Their petition was strongly supported, amongst others, by the renowned chemist and physician Heinrich Ranke. On the opposite side, however, they faced a formidable adversary: Max von Pettenkofer. Pettenkofer not only denied the definite link between polluted drinking water supplies and epidemic disease. He also argued that the Isar's self-purificatory powers were strong enough to handle the Munich's sewage without difficulty. Dilution, Pettenkofer argued, readily dissolved organic matter, while the river water also effectively neutralised the harmful potential of germs.[82] Given that the Government of India had resorted to Pettenkofer for decades to defend its public health policy, the lack of reference to the Munich controversy is rather surprising. The only official who built his opinion – even though only partly and indirectly – on Pettenkofer's work was D.W. Aikin, the sanitary engineer responsible for the chemical water analyses at Kanpur and Lucknow in 1892. Basic for his belief in the germicidal agency of river self-purification was Alessandro Serafini's study on the Isar, which the Italian scientist conducted as part of Max von Pettenkofer's team at the Institute for Hygiene in Munich.[83]

There are a number of possible reasons for this strong metropolitan focus. For one, Britain maintained a leading position in matters of sewerage technology, sewage treatment, river pollution and water quality research for most of the nineteenth century, and European and American scientists and administrators generally looked towards Britain for guidance. Britain's dominance in these fields diminished only from the late 1880s onwards with the advent of biological sewage treatment, for which the work of US American scientists and engineers in particular was crucial.[84] On a more general level, colonial science remained quite closed to international scientific discourse until the late nineteenth century. Nevertheless, Britain functioned as the major reference point for colonial debates on sewage disposal and river pollution even after the introduction of biological sewage treatment, as the following chapters will show.

Notes

1 R. Smeaton, Secy GoNWP, 1.5.1890, UPSA, GoNWP, Municipal Dpt., File 3/63/1890, Box 532, 'Benares Water-Supply and Drainage Project', Prgs June 1890, Notes and Orders, p. 2.
2 See, for example: Arnold 1993; Harrison 1994; Kumar 2001; Pati and Harrison 2006; Pati and Harrison 2009a; Chakrabarti 2012. For a recent state of the art, see Pati and Harrison 2009b.
3 Sharan 2011: 427. Among the few exceptions are Mann 2007; Prashad 2001; and Arnold 2013.
4 G. Hutcheson, Sanitary Commissioner NWP, to Secy GoNWP, 22.3.1890, UPSA, GoNWP, Municipal Dpt., File 3/63/1890, Box 532, 'Benares Water-Supply and Drainage Project'.
5 G. Hutcheson, 'Memorandum on the sewage pollution of rivers, with special reference to the rivers of the North-Western Provinces and Oudh', n.d. [September 1890], NAI, GOI, Home Dpt., Sanitary Branch, Prgs April 1893, Nos 17–25.
6 G. Hutcheson, Sanitary Commissioner NWP, to Secy GoNWP, 22.3.1890, UPSA, GoNWP, Municipal Dpt., File 3/63/1890, Box 532, 'Benares Water-Supply and Drainage Project'.
7 G. Hutcheson, 'Memorandum on the sewage pollution of rivers, with special reference to the rivers of the North-Western Provinces and Oudh', n.d. [September 1890], NAI, GOI, Home Dpt., Sanitary Branch, Prgs April 1893, Nos 17–25.
8 Ibid.
9 Ibid. (emphasis mine).
10 From the mid-nineteenth century onwards, the Indian government employed medical officers as chemical analysts, with the task to examine different sources of water supply and to advise local authorities on sources for drinking water. This was part of a wider repertoire, in which the chemical analyst was not only concerned with water but checked a number of substances for contamination, chiefly for toxicological purposes and in the context of medical trials (Sharan 2011: 429).
11 W. Venis, Analyst to the Benares Municipality, to J. White, Chairman, Banaras Municipal Board, 15.9.1890, UPSA, GoNWP, Municipal Dpt., File 491A/4, Boxes 19 & 20, 'Pollution of rivers by the discharge of sewage; Benares water works and drainage'.
12 Other experts and books of the pre-bacteriological era referred to were, among others, Edmund Parkes' *Manual of Hygiene* and George Wilson's *A Handbook of Hygiene and Sanitary Science* (1873). Wilson was medical officer of health for mid-Warwickshire in England (see Hamlin 1987: 441, 445, 448). In Britain, public health textbooks adequately covering bacteriology were published only from the late 1890s onwards (Worboys 2000: 239).
13 Harrison 1994: 209–10.
14 Buck's *A Treatise on Hygiene and Public Health*, first published in 1879, was a widely used reference work during the late nineteenth century (Meckel 2001: 76).
15 'From the Honorary Secretary, Public Health Society of Calcutta, to the Secretary to the Government of India', 29.7.1890, NAI, GOI, Home Dpt., Sanitary Branch, Prgs August 1890, No. 44. The miasmatic theory seems to still have held some influence in the Society, as it warned that the 'exhalations from streams charged with sewage' were a source of danger and could destroy aquatic animal life (ibid.).
16 Ibid.; G. Hutcheson, Sanitary Commissioner NWP, to Secy GoNWP, 22.3.1890, UPSA, GoNWP, Municipal Dpt., File 3/63/1890, Box 532, 'Benares Water-Supply and Drainage Project'.
17 A.J. Hughes, 'Note on the disposal of sewage at Benares into the Ganges', n.d. [February 1890], UPSA, GoNWP, Municipal Dpt., File 3/63/1890, Box 532, 'Benares Water-Supply and Drainage Project'. Hughes here specifically points to the research

of one 'Nicholls Galton'. Quite likely, he meant John Nichol, who in a study on the Mississippi noted that

> with its rapid current, [the Mississippi] will quickly carry away the discharge of the sewers, and from the quantity and nature of the river water in constant and rapid motion, will quickly dilute and deodorize the sewage, so that all traces of it will disappear in a short distance below the port of discharge.
>
> (quot. in Tarr 1996: 149)

18 A.J. Hughes, Supervising Engineer, Municipal Water-Works, NWP, to Chairman, Banaras Municipal Board, 2.9.1890, UPSA, GoNWP, Municipal Dpt., File 491A/4, Boxes 19 & 20, 'Pollution of rivers by the discharge of sewage; Benares water works and drainage'.

19 William Santo Crimp (1853–1901) was a widely recognised English sanitary engineer (w.a. 1901a: 458; w.a. 1901b: 343–6). During the later stage of his career, he became directly involved with several sewerage projects in India, a matter which will be discussed further down.

20 A.J. Hughes to Chairman, Banaras Municipal Board, 2.9.1890, UPSA, GoNWP, Municipal Dpt., File 491A/4, Boxes 19 & 20, 'Pollution of rivers by the discharge of sewage; Benares water works and drainage'.

21 Ibid.

22 Ibid.

23 G. Hutcheson, Sanitary Commissioner NWP, to Secy GoNWP, 22.3.1890, UPSA, GoNWP, Municipal Dpt., File 3/63/1890, Box 532, 'Benares Water-Supply and Drainage Project'; H. Maude, Jr. Secy GoPun, to Secy GoNWP, 10.9.1890, UPSA, GoNWP, Municipal Dpt., File 491A/4, Boxes 19 & 20, 'Pollution of rivers by the discharge of sewage; Benares water works and drainage'.

24 'Note of a conference held at Delhi on the 15th November 1890, on the question of putting the sewage of Delhi into the river Jumna', UPSA, GoNWP, Municipal Dpt., File 491A/4, Boxes 19 & 20, 'Pollution of rivers by the discharge of sewage; Benares water works and drainage'.

25 APAC, IOR, P/4297, GoNWP, Municipal Dpt., Prgs May 1893 (B), No. 6.

26 A.J. Hughes to Chairman, Banaras Municipal Board, 2.9.1890, UPSA, GoNWP, Municipal Dpt., File 491A/4, Boxes 19 & 20, 'Pollution of rivers by the discharge of sewage; Benares water works and drainage'.

27 Together with the sanitary commissioners, inspector-generals (also called surgeon-generals) of civil hospitals acted as the principal advisors on public health matters to the provincial governments (Mushtaq 2009: 7).

28 J. Richardson, 'Memorandum', n.d. [22.4.1890], UPSA, GoNWP, Municipal Dpt., File 3/63/1890, Box 532, 'Benares Water-Supply and Drainage Project', Prgs June 1890, Notes and Orders, p. 5.

29 J. Richardson to G. Hutcheson, 16.9.1890, NAI, GOI, Home Dpt., Sanitary Branch, Prgs April 1893, Nos 17–25.

30 Isaacs 1998: 296.

31 Worboys 2000: 247–54, 265.

32 A.J. Hughes to Chairman, Banaras Municipal Board, 2.9.1890, UPSA, GoNWP, Municipal Dpt., File 491A/4, Boxes 19 & 20, 'Pollution of rivers by the discharge of sewage; Benares water works and drainage'.

33 Ibid. The passages highlighted by Hughes refer, among others, to the difficulties bacteriologists still faced when trying to distinguish between different species of disease germs, and their difficulties in fulfilling Koch's four postulates in the case of abdominal typhus, ague, leprosy, diphtheria and cholera.

34 R. Smeaton, Secy GoNWP, 1.5.1890, UPSA, GoNWP, Municipal Dpt., File 3/63/1890, Box 532, 'Benares Water-Supply and Drainage Project', Prgs June 1890, Notes and Orders, p. 5. The secretary commended Richardson's advice as 'clear [...]

and succinct [...]' and found it 'satisfactory that, coming as it does from an able and outspoken officer, it coincides with what I understand to be the views of the Government on the subject' (ibid.).

35 *British Medical Journal*, 23 August 1913, p. 526.

36 J. Richardson to Secy GoNWP, 18.9.1890, NAI, GOI, Home Dpt., Sanitary Branch, Prgs April 1893, Nos 17–25. Here it is important to note that none of these riparian towns possessed a *sewerage* system at the time, i.e. excreta were still removed by hand and trenched, even though some amount would have accidentally found its way into the rivers through the surface drains. Therefore, it can be rightly assumed that what was actually being discharged through the drains was a mix of street drainage and a variety of wastewaters, including that of households and manufactures.

37 R. Smeaton, Secy GoNWP, to Commissioners of Agra, Allahabad and Banaras Divisions, 11.11.1890, NAI, GOI, Home Dpt., Sanitary, Prgs April 1893, Nos 17–25. *Kanungos* were part of the revenue administration and acted as accountants or field inspectors (Goswami 2004: 138–9).

38 J. Richardson, Inspector-General of Civil Hospitals NWP, to Secy GoNWP, 23.9.1891, NAI, GOI, Home Dpt., Sanitary, Prgs April 1893, Nos 17–25.

39 This is evident from the fact that the discussion ends with the submission of Richardson's evaluation.

40 NAI, GOI, Home Dpt., Sanitary, Prgs April 1893, Nos 17–25. In contrast, people at Agra generally resorted to wells, considering Yamuna water unwholesome (no indications are given as to why). At Allahabad, too, Ganges water was much preferred over that of the Yamuna.

41 Goddard 2004; Arnold 2000: 133, 151–2; Voelcker 1893.

42 R. Smeaton, 'Minutes of Meeting', 14.5.1890, UPSA, Go NWP, Municipal Dpt., File 3/63/1890, Box 532, 'Benares Water-Supply and Drainage Project', Prgs June 1890, Notes and Orders, p. 6.

43 Baldwin Latham (1836–1917), as mentioned previously, was one of the most renowned sanitary engineers of his days, and by 1868 had designed the sewerage, irrigation and waterworks for 15 English towns. His manual *Sanitary Engineering* on the construction of sewerage and house drainage, first published in 1873, was regarded as a classic (Insley 2004). As Christopher Hamlin notes, it was very common for sanitary engineers in Victorian Britain to speak authoritatively on epidemiology and the aetiology of waterborne diseases (Hamlin 1987: 57).

44 A.J. Hughes to Baldwin Latham, 26.5.1890, APAC, IOR, P/3598, GoNWP, Municipal Dpt., Prgs December 1890, No. 109.

45 B. Latham to A.J. Hughes, 5.8.1890 and 18.7.1890, APAC, IOR, P/3598, GoNWP, Municipal Dpt., Prgs December 1890, Nos 105 and 107.

46 Carkeet James 1906a: 252–3, 268–9, 312–13.

47 Quot. in Colvin (1894: 522).

48 Latham 1878: 72–3. The same paragraph was reprinted in the American edition of the book published in 1884 (Latham 1884: 7). Significantly, this passage had been used by the sanitary commissioner of India, W.R. Rice, to make his case for sewage farms (see Chapter 2).

49 B. Latham to A.J. Hughes, 5.8.1890, APAC, IOR, P/3598, GoNWP, Municipal Dpt., Prgs December 1890, No. 107.

50 W. Venis, Analyst to the Banaras Municipality, to J. White, Chairman, Banaras Municipal Board, 15.9.1890, UPSA, GoNWP, Municipal Dpt., File 491/4, Boxes 19 and 20, 'Pollution of rivers by the discharge of sewage; Benares water works and drainage', Prgs December 1890, No. 35; G. Hutcheson, 'Memorandum', [September 1890].

51 G. Hutcheson, 'Memorandum on the sewage pollution of rivers, with special reference to the rivers of the North-Western Provinces and Oudh', n.d. [September 1890], NAI, GOI, Home Dpt., Sanitary Branch, Prgs April 1893, Nos 17–25.

52 It has to be noted here that during the early 1890s, bacteriological water analysis itself was still very much of an evolving discipline. In Britain, the influence of bacteriology on water analysis remained very weak for many years. Manuals for public health officials came to include bacteriological techniques, but the majority of British analysts stuck to chemical methods. Some of them even ignored bacteriology completely. For many years, bacteriological water analysis moreover remained confined to counting the number of colonies growing on a gelatine-coated plate, and even this was fraught with ambiguities and difficulties. It was only by 1895, thanks to the developments in determinative bacteriology, that water could be analysed for specific pathogens. Still, uncertainty and contradictions prevailed, and water analysis continued to be a highly controversial field. As mentioned previously, the advent of bacteriology did not help to clarify existing confusions with regard to water quality well into the 1900s (Hamlin 1990: 270–7; Hamlin 1987: 454–5, 462).

53 D.W. Aikman, 15.3.1892, UPSA, GoNWP, Municipal Dpt., File 491/4, Boxes 19 and 20, 'Pollution of rivers by the discharge of sewage; Benares water works and drainage', Prgs September 1892, No. 26(a).

54 D.W. Aikman, 15.3.1892, UPSA, GoNWP, Municipal Dpt., File 491/4, Boxes 19 and 20, 'Pollution of rivers by the discharge of sewage; Benares water works and drainage', Prgs September 1892, No. 26(b).

55 Quot. ibid. See also Serafini's original text (Serafini 1891: 354). Serafini (1859–1911) worked at the institute for two years, after which he took office as professor of hygiene at the University of Padua (D'Orazio 1998: 298).

56 D.W. Aikman, 15.3.1892, UPSA, Go NWP, Municipal Dpt., File 491/4, Boxes 19 & 20, 'Pollution of rivers by the discharge of sewage; Benares water works and drainage', Prgs September 1892, No. 26(b).

57 C. Perrin, Sanitary Engineer NWP, to Secy GoNWP, 24.10.1893, UPSA, GoNWP, Municipal Dpt., File 491/4, Boxes 19 & 20, 'Pollution of rivers by the discharge of sewage; Benares water works and drainage', Prgs December 1893, No. 22; C. Perrin, Sanitary Engineer NWP, to Secy GoNWP, 22.9.1894, ibid., Prgs October 1894, No. 1.

58 J.S. Beresford, Supervising Engineer, Municipal Works, NWP, to Secy GoNWP, 5.10.1892, UPSA, GoNWP, Municipal Dpt., File 491/4, Boxes 19 & 20, 'Pollution of rivers by the discharge of sewage; Benares water works and drainage', Prgs November 1892, No. 1.

59 Colvin 1894: 517.

60 Hewlett 1939.

61 Chakrabarti 2012: 45; Arnold 2000: 142.

62 E.H. Hankin to Secy GoNWP, 19.11.1892, UPSA, GoNWP Municipal Dpt., File 491/4, Boxes 19 & 20, 'Pollution of rivers by the discharge of sewage; Benares water works and drainage', Prgs November 1892, No. 5.

63 *Annual Report of the Chemical Examiner and Bacteriologist to the North-Western Provinces and Oudh and the Central Provinces, 1894*, pp. 5–6; and *Annual Report of the Chemical Examiner and Bacteriologist to the North-Western Provinces and Oudh and the Central Provinces, 1895*, pp. 41–8.

64 Hankin 1896.

65 As it will be remembered, this argument was used by the inspector-general of civil hospitals J. Richardson to challenge sanitary commissioner G. Hutcheson a few years earlier.

66 Hankin 1896: 512.

67 Ibid.: 520. A second experiment conducted with cholera microbes in Ganges water showed no germicidal action. Hankin explained this with the long time lapse of 50 to 60 hours which it took to transport the samples by train from their source to Agra.

68 Ibid.: 522.

69 *Annual Report of the Chemical Examiner and Bacteriologist to the North-Western Provinces and Oudh and the Central Provinces, 1894*, p. 6; *Annual Report of the*

Chemical Examiner and Bacteriologist to the North-Western Provinces and Oudh and the Central Provinces, 1895, pp. 45–6. The sources consulted for this book do not suggest that Hankin continued his work on the self-purification of Indian rivers. In 1896, he was appointed as a member of the Plague Research Committee with the task to investigate the vitality and longevity of the plague bacillus under different conditions (Harrison 1994: 152), an office which is likely to have left him with little time to continue his earlier research.

70 Hamlin 1988a: 122.
71 Driver and Martins 2005; Arnold 2006 (esp. p. 111); Arnold 1996: 5–10.
72 Chakrabarti 2012: 3–5.
73 See Chapter 2.
74 Chakrabarti 2012: 9–13, 25.
75 S.J. Thomson, Sanitary Commissioner NWP, to Secy GoNWP, 14.2.1898, APAC, IOR, P/5419, GOI, Home Dpt., Municipalities, Prgs May 1898, Nos 8–11.
76 Sharan 2011: 429.
77 Kumar 2013: 91–2. There are no signs that the North-Western Provinces consulted either Cunningham or Lingard.
78 Arnold 2000: 135 (quote ibid.).
79 MacLeod 1975: 352–4; Kumar 2013: 105.
80 None of the major works on British river pollution controversies suggest Voelcker's involvement. See, for example: Hamlin 1987; Hamlin 1990; Hamlin 1988a; Breeze 1993; Luckin 1986; Rosenthal 2014. Voelcker is not to be confused with his father, the agricultural chemist Augustus Voelcker (1822–84), whose opinion was consulted by the first British Rivers Pollution Commission during the 1860s (Hamlin 1987: 273).
81 w.a. 1937.
82 Pettenkofer 1890; Pettenkofer 1891; *Deutsche Vierteljahrsschrift für öffentliche Gesundheitspflege*, Vol. 24 (1892), 108–35.
83 Serafini 1891.
84 Steinberg 1991: 214–15; Melosi 2008: 109–10; Mohajeri 2005: 73; Goubert 1989: 96.

3 Local self-government and river pollution

The two previous chapters have examined the Government of India's and the North-Western Provinces' approaches towards the organised discharge of sewage into the Ganges and other Indian rivers. This chapter zooms in to the 'ground-level' of the city, asking what positions were adopted by municipal governments and urban populations, and to what extent they attempted to and were able to influence the choice of sewerage technology. On the background of the Ganges's central role in Hindu religion, it is appropriate to include a specific focus on Hindu municipal commissioners and citizens in this context, i.e. to investigate in what way religious perspectives on the river shaped their viewpoints and actions, and to what extent these perspectives influenced the outcome of government policy.

As mentioned previously, Indian municipalities from the 1870s onwards carried the financial responsibility for urban sanitation, and the funds for waterworks and sewerage systems had to be raised primarily via municipal tax revenues.[1] Scholars writing on the development of public health in British India have generally agreed that the success of sanitary reform was very limited.[2] As numerous studies have demonstrated, urban waterworks, drainage and sewerage infrastructures remained piecemeal and often clearly prioritised European quarters.[3] Explanations for this state of things have however varied. Early studies on the impact of local self-government on the development of public health in British India took on rather one-sided perspectives, either claiming that British colonial officials actively curbed initiatives taken by Indian municipal commissioners towards sanitary reform,[4] or asserting, in contrast, that public health services were introduced only where British officials fostered them, while Indians overall remained indifferent towards Western concepts of sanitation.[5] More recent studies have presented more nuanced analyses, demonstrating that the direction of public health in any given area was influenced by a number of different forces. As Mark Harrison has put it, 'the development of sanitation depended on a dynamic interaction between government, the local revenue situation, municipal commissions, sanitary officers, and the indigenous population'.[6] An overview of different case studies[7] suggests that there were two major determinants: First, the colonial state's 'fiscal conservatism'[8] left municipalities dramatically underfunded. Generally, the bulk of colonial revenues was spent for the military,[9] the home charges (including, among others, the pensions of British colonial officials) and to cover the 'India debt'. To the state, urban planning and development were of little interest, unless a city held strategic and/or commercial importance, and was home to a significant number of Europeans. In the wake of fiscal decentralisation in the 1870s, municipalities found themselves with the financial responsibility for urban sanitary infrastructures, but were left with no means to raise adequate funds. In the North-Western Provinces and elsewhere, the main source of municipal income was the octroi tax, which rendered municipal incomes sensitive to economic fluctuations and made it hard to budget for sanitary expenditure. The only way to finance expensive large-scale infrastructure projects such as waterworks and sewerage systems was to introduce new taxation and to raise loans or grants-in-aid from the central and provincial governments. However, new taxation was prone to provoke violent protests from urban populations, while the central and provincial governments' willingness to issue loans and grants-in-aid as requested by municipalities varied.[10]

As to Indian attitudes towards public health, the second major determinant, there existed great variations not only from place to place, but also within municipalities. While some Indian commissioners actively embraced Western concepts of public health and campaigned for the adoption of waterworks and other sanitary measures and infrastructures, others remained overall indifferent or openly antagonistic to these for religious, cultural, or financial reasons. Indian opinion generally weighed more strongly after the enlargement of the representative element in municipal bodies in 1882.[11] Apart from colonial fiscal conservatism and Indian attitudes towards public health, a number of other factors have been cited for the failure of urban sanitary infrastructures, such as non-cooperative administrative bodies on different levels, paralysing and blocking each other's work,[12] or the fact that Western technologies were often inherently unsuitable for Indian circumstances, being too cost- and resource-intensive.[13]

The case studies of Banaras and Kanpur presented in this chapter demonstrate that the colonial state's fiscal conservatism was the major guiding principle in determining the municipalities' choice of sewerage technology. The Banaras and Kanpur Municipal Boards both initially made sewage farms a constitutive part of their sewerage projects, a part which in Banaras was strongly insisted upon by many Hindu citizens, including most Hindu municipal commissioners. However, soon after taking up loans for waterworks construction, the Boards found themselves in a precarious financial situation, and with no help forthcoming from the central and provincial authorities, they were forced to drop the idea of a sewage farm indefinitely. The Kanpur Municipal Board even had to abandon its sewerage project as a whole in the end.

The detailed documentation of the Banaras case allows us not only to reconstruct the practical working of colonial fiscal conservatism and the attitudes of Indian municipal commissioners, it also facilitates an assessment of the less direct, but potentially no less powerful role assumed by the citizens of Banaras. It was on them that the power of the elected segment in municipal bodies relied (even though actual voting power was restricted to a thin layer of upper-class residents),[14] and they could put considerable pressure on administrators by resorting to public protest. The riots against a new house tax in Banaras in 1810/11 present an early example of such protest against government intervention.[15] The discharge of a ritually impure substance such as excreta-containing sewage into the Ganges was potentially highly controversial from a Hindu religious point of view, and it was possibly even more so in Banaras, a *tīrtha* holding special eminence in Hindu sacred geography. Therefore, this chapter takes a closer look at incidences of popular unrest in Banaras during the early 1890s and tries to assess in how far they were directed at the proposed discharge of sewage into the Ganges.

On 19 November 1886, the Kashi Ganga Prasadini Sabha held its inaugural meeting in the town hall of Banaras. A few months earlier, the Sabha's leading members had approached the sanitary commissioner of the North-Western Provinces with a number of suggestions for sanitary improvements in the city. The pollution of the Ganges and the bathing *ghāṭs* through the old city drains was of particular concern, and during its inaugural meeting, the Sabha declared it as its aim 'to concert measures to prevent the pollution of the Ganges, within the limits of the sacred city of Benares, by the discharge of filth into it through the sewers'.[16] The inaugural speech of member Pramada Das Mittra elaborated this further:

> [Pramada Das Mittra] deprecated from a religious as well as a sanitary standpoint, the discharge of filth into the river, which, sacred as it was everywhere in the eyes of the pious Hindu, attained tenfold sanctity by contact with Kasi (Benares) the holiest of cities. He said that at the time when the present drains were constructed to discharge sewage at the very places of ablution and prayer, it was a matter of necessity and could not be helped, but now when Western wisdom renders it perfectly feasible to rapidly remove by means of scientifically constructed drains and pipes the filth of the town beyond its limits, it would be sinful on the part of the citizens not to vindicate their reverence for their sacred city by exerting themselves to their utmost to accomplish this object. [...] Would not that philanthropy and religious zeal [...] which constructed the noble ghâts and temples for the temporal and spiritual benefit of the people, be called forth in a cause no less sacred and beneficent and show themselves in the Ganga of Kasi flowing free of all foul matter?[17]

Mittra's speech reveals two important aspects. First, the religious argument clearly outweighs the secular call for sanitation. The primary objective of applying the new Western technology was to keep the sacred city and its places of prayer, the *ghāṭs*, free from filth, and thus in a condition worthy of their spiritual significance. Secondly, the aim of the Sabha was not to prevent the pollution of the Ganges per se, but to prevent the pollution of the Ganges within the sacred zone of Kashi. For the Sabha, the cleanness of the *tīrtha*, and by extension of the 'Ganga of Kasi', assumed primary importance over the cleanness of the Ganges as a whole.

The local press reported favourably on the Sabha's meeting, and expressed its hopes that the course of the drains may be diverted from the city as soon as possible.[18] The Sabha's declared support for a municipal sewerage project was moreover warmly welcomed by the provincial authorities. Sanitary commissioner Planck expressed his satisfaction about the 'considerable awakening, in regard to the advantages and necessity of sanitary improvement' shown by the Banaras elite. And Sir Auckland Colvin accepted the Sabha's invitation to preside over it as a patron as a sign of support.[19] However, the Sabha's religious framework caused some apprehensions early on. Bireshwar Mittra, nominated member and secretary to the Banaras Municipal Board, and a zealous supporter of the waterworks and sewerage projects, expressed his doubts about the successful cooperation between the municipal administration and the Sabha. He warned that since the aim of the Sabha was 'semi-religious [...], *i.e.* to prevent the pollution of the river Ganges within the limits of the sacred city', it was 'possible that the most feasible project of sewerage which the Board may have to adopt be not in exact coincidence with the views of the Sabha'. In Mittra's view, therefore, the municipal board was not to rely too much on the Sabha's help.[20]

The Banaras municipal board was established in 1884 (replacing the earlier municipal committee founded in 1867) and featured a considerable number of elected members. In 1888, eighteen were elected and seven nominated, the latter including the British magistrate who was also its chairman.[21] The Indian commissioners hailed from the social and economic elite of the city, including aristocrats, merchant-bankers and others.[22] Significantly, the leading members of the Kashi Ganga Prasadini Sabha were simultaneously members of the board. Bireshwar Mittra, for instance, was the Sabha's joint secretary, a fact about which he himself felt somewhat uncomfortable. Believing it unworthy of a rich city such as Banaras to rely on the charity of the public for sanitary improvement, he admitted that he was not 'able fully to sympathize with the object of the Sabha'.[23]

Spurred by the Sabha, and under considerable pressure from its British chairman and the provincial government, the Banaras municipal board in 1888 earnestly reconsidered the proposals for waterworks and sewerage. Lieutenant-Governor Sir Auckland Colvin had made it clear to the board that, following the Indian government's resolution of the previous year, he was determined to have waterworks introduced in all of the province's major towns.[24] While the majority of the board were unanimous on the necessity of waterworks and sewerage, the projected financial outlay gave rise to hesitation. The main source of income to the municipality was the octroi tax, and it was clear that any large scheme would involve the introduction of new taxes and the raise of a considerable loan. Moreover, several commissioners felt that they had not been given enough time to consider the details of Hughes' project. Much to the discontent of their chairman, who in a hardly concealed attempt at intimidation warned them that 'Government would have to be apprized of the fact that the Board refuse to do their duty and that other measures must be taken', they succeeded in setting up a subcommittee to look at the matter more closely. After a good deal of deliberation, the sub-committee and the board accepted Hughes' proposals in March 1889. The total cost of the projects was estimated to almost 40 lakhs rupees, 22 lakhs for the waterworks and 16.5 lakhs for sewerage.[25] Following Hughes' estimate, Bireshwar Mittra drew up a financing plan, to which the board also agreed. Funds were to be incurred by a loan from the provincial government, and by introducing new and raising existing taxation. New taxes were to include a residence tax, a drainage tax and the water rates, all applying to houses lying with the area connected to the new services. Mittra was hopeful that the 10 lakhs provided for by the residence tax could be quickly paid off by private subscriptions collected through the Kashi Ganga Prasadini Sabha and others. Additional to the afore-mentioned taxes, the octroi tax was to be raised for a number of items already taxed, and extended to others, such as leather and glass goods. Also, a wheel tax was to be introduced for carriages, horses and bullock carts.[26]

Before long, the municipal commissioners faced a storm of protest from the townsfolk. Under the leadership of the Sujan Samaj, an organisation of Western-educated pleaders and clerks,[27] a crowd of several thousand people gathered in front of the town hall on April 27 to protest against the board's acceptance of Hughes' project and the prospects of enhanced taxation. According to the protest leaders, the municipal board had acted rashly and without consulting the people of their wards whom they represented. Its decisions were therefore entirely rejected. Tied to this was a general reproach as to the alleged lack of transparency with which the board conducted its businesses. It was resolved that a committee of inquiry arranged by the protesters should be granted four months to understand and assess the particulars of Hughes' project and its practical and financial implications, and that no additional taxes should be raised until the matter was settled.[28] The proceedings of the protest meeting, together with a petition, were sent to the government of the North-Western Provinces. Sir Auckland Colvin, however, gave the municipal board his full backing. In his reply, he agreed that the committee should be furnished with all the required information, and guaranteed that its criticisms would receive due attention. But at the same time he unambiguously pointed out that to abandon the waterworks and sewerage schemes was completely out of question. Banaras, Auckland Colvin emphasised, was one of the 'foulest cities in the Upper Provinces' and a 'permanent source of infection to the whole peninsula'. Its sanitary improvement, therefore, was of much more than mere local importance, and it would not be permitted that the whole of the North-Western Provinces were 'exposed to infection from Benares, solely from consideration to Benares rate-payers'.[29]

The meeting's official proceedings and the associated petition, as well as the contents of numerous other petitions sent to government by Banaras citizens and representatives of a variety of professions,[30] reveal that the main motive underlying the protests was economic. Increased taxation and the raise of a government loan, citizens feared, would put an unprecedented financial burden on them. However, other sources suggest that religious motives were also involved to some extent. According to an article published in the *Pioneer*, some of those attending the meeting stressed that the Hindu inhabitants of Banaras would never drink pipe-water on religious grounds, and would refuse to do so even if pressurised by municipal by-laws and government orders. A look at Bireshwar Mittra's memorandum of August 1888 shows that this issue had already been smouldering for a while. Writing about the waterworks, Mittra scathingly referred to the opposition from certain 'bigoted Hindus', found even among the 'leaders of public opinion', whose 'senseless prejudice' and 'superstition' rendered them antagonistic to piped water supplies, perceiving these as an attack on their religion. However, he was confident that opposition would soon vanish once the works were in operation, much as it had been in Calcutta many years before.[31] Moreover, the *Pioneer* reported, it was alleged that the Kashi Ganga Prasadini Sabha had been misled by the board regarding the disposal of the sewage, which, 'under Mr Hughes' scheme, would be thrown into the sacred river, while the object of this Society was *expressly* to prevent this'.[32] No document seems to be available that talks about the immediate reaction of the Sabha itself to the proposed temporary discharge of sewage into the Ganges within the sacred zone of Kashi. The Sabha temporarily disintegrated in 1889 (see below), and it is likely that during April it officially did not exist at all. According to Hughes, the majority in the municipal board strongly favoured the disposal of the sewage on a sewage farm and were 'most unwilling' to have it put into the Ganges at all, as it was 'strongly against Hindu prejudices'. Nevertheless, they came round to his 'practical' view that a sewage farm at the present stage was not feasible due to a lack of funds, as well as the absence of information about the quantity and quality of the sewage to be dealt with.[33] As most of the leading members of the Sabha were at the same time municipal commissioners, it may be assumed that they similarly had come to favour a sewage farm but given in to Hughes' 'practical view' for the time being. In any case, the provincial government was hardly sympathetic to any other demands. Should the Sabha consider the discharge of the sewage into the Ganges above the Varana inappropriate, one official remarked laconically, it would most certainly be ready to pay the four lakhs extra-cost for the transportation of the sewage over the Varana.[34]

In view of the above, it seems safe to conclude that the popular protests vented on April 27 were primarily based on financial concerns, while involving some religious element. A look at the coverage of the event in the local press similarly suggests this. The *Hindustan* approved of the sewerage project, but supported the criticism and demands expressed at the protest meeting with regard to financing and the lack of consideration for public opinion. The *Bharat Jiwan* expressed its hopes that government may look into the matter seriously, as the new taxes and the raise in octroi would be a source of great pain to the people.[35]

In March 1890, the municipality put its new taxation scheme into force. Popular feeling, unsurprisingly, rose again. As the *Rafi-ul-Akbar* observed, 'the late increase in municipal taxation at Benares [was] generally viewed with disfavour by the poorer classes of traders and shopkeepers' and the butchers had already struck work. Other classes were expected to follow suit.[36] The tax raise had come at a most inauspicious moment, since people in the North-Western

Provinces were already suffering considerable hardships due to a grain shortage. In Banaras, it was the poor class of Muslim weavers that was hit worst, as increasing grain prices caused a drastic decline in the demand for their products.[37] On these backgrounds, Sir Auckland Colvin's ambitious undertakings were severely criticised by the Indian press. 'No doubt a pure and plentiful supply of water is a very desirable thing', the Lucknow-based newspaper *Hindustani* wrote, but the way in which the measure was carried out in Banaras and other cities threatened to cause ruin to all but the wealthy.

> [I]f Sir Auckland Colvin paid a visit to [Banaras] at night, His Honor would hear the cries of the people groaning under taxation. They already pay the octroi duty, and are now threatened with a house tax; and it has been proposed to levy a duty on pilgrims. The municipality will borrow 40 lakhs of rupees at an interest of 5 per cent. The interest on the loan will amount to 1.75 lakh a year, and the annual expenses of maintenance of the waterworks will be over 1.5 lakh. The present municipal revenue being 1.5 lakh a year, the municipality will have to raise them to 6 lakhs in order to be able to maintain the waterworks, pay the interest, and clear the loan in 40 years! [...] It is difficult to understand how the people will be able to bear so heavy a burden.[38]

For Kanpur and Agra, the newspaper predicted similar hardships.

In 1891, popular unrest in Banaras reached its climax. That year, it turned out that the construction of the waterworks pumping station at Bhadaini threatened the existence of a small temple located close to the construction site. The municipal board was careful enough not to actively demolish the temple, but instead decided to wait for it to crumble down by itself as the ground around it kept being excavated. In April, the board held a meeting on the issue, during which a crowd of around 500 people gathered. Once the meeting ended, a major commotion ensued, with 5,000 to 6,000 people assembling at the temple site and raising cries to 'destroy the machinery!'. Significantly, this crowd included different classes of Hindus, as well as Muslim weavers. In the violent eruptions that followed, protesters threw boilers and pumps into the river, tore up supply pipes, and pitched big pieces of iron well curb into the culling. From there, they proceeded through the streets of Banaras, destroying street lamps, cutting down telegraph wires, and looting and wrecking the telegraph office. The railway station, too, was looted. Only thanks to massive police efforts, supported by soldiers of a nearby European regiment, could the tumult be crushed.[39]

In her assessment of these events, Sandria B. Freitag has suggested that rioters 'particularly targeted symbols of imperially imported technology', and that 'a protest that had begun over a temple had been rerouted against the technology of the state'.[40] While it is certainly true that objects of Western technology were the protesters' prime target (another target was the house of a particularly unpopular Indian municipal commissioner), it is important not to misinterpret this as an expression of an inherent repudiation of Western technology.[41] Instead, I contend that these protests were aimed at the *way* in which the colonial state tried to enforce the waterworks and sewerage projects, which held little regard for the economic situation of the general population and religious sensibilities. Thus, the protests have to be viewed as an outlet for more general political discontent.

In sum, the 1891 protest represents a continuation of the protest staged at the town hall two years earlier. At the heart of both lay economic concerns, while religious motives played a secondary, catalysing role. The proposed discharge of sewage into the Ganges in particular is mentioned nowhere in 1891. This may also be explained with the fact that sewer construction had not begun yet, and thus the immediate attention rested on the waterworks and the temple they threatened. What both the protests and their contexts clearly illustrate is the pressure under which municipal boards found themselves when aiming at large-scale urban infrastructure projects: Not only did they enter a tremendous financial commitment sure to weigh heavily on municipal finances for many decades to come (a commitment which was literally forced upon the Banaras municipal board by Sir Auckland Colvin), they also faced intense pressure from their citizens when struggling to raise more municipal income.[42]

Bengal's objections against the projected discharge of untreated sewage into the Ganges put a second serious challenge before the Banaras municipal board. In April 1890, they were communicated to the chairman and laid in front of the board's drainage and waterworks committee for consideration and report. In September, the committee resolved that in view of the great weight of the question the board be recommended to declare itself incapable of taking any decision, and to leave the matter entirely in the hands of the provincial government.[43] The latter, however, refused. In a letter to the board, Sir Auckland Colvin underlined the commissioners' responsibility to form their own opinion on municipal matters. At the same time, he undertook a rather obvious effort to lead the board into the desired direction, by stressing the imminent threat to the financial stability of the municipality. If the board left criticisms unanswered, the Lieutenant-Governor warned, they would either have to abandon the sewerage project completely, or incur an unnecessary burden of expenditure by adding an 'elaborate system of sewage purification at a very great cost'. The note drawn up by A.J. Hughes,[44] he pointed out, contained all the information needed by the board and should, together with further expert opinion arranged by Hughes, be used by them to fortify their position. The expenditure incurred by engaging expert opinion would be inconsiderable in comparison to the costs incurred by purification measures.[45] The board, however, remained adamant. The majority of the members were of the opinion that English experts were unfamiliar with Indian conditions (e.g. the fact that the corpses of cholera and smallpox victims were thrown into rivers) and that therefore their expertise was of little use. Any expert's opinion could be disproved by another, they insisted, and the money invested in such a procedure would be wasted. At the same time, proposals for the appointment of a committee in India were also refuted. Some commissioners contended that the whole issue was irrelevant, since rivers had always been used as waste carriers in Banaras.[46] Confronted with the board's refusal to arrange for experts, the chairman suggested that Baldwin Latham, who was expected to arrive in India shortly, be asked to not only report on the technicalities of the Banaras and Kanpur sewerage systems, but also on the sewage disposal question.[47]

In 1891, preliminary steps for the construction of the sewers were taken up.[48] Between 1892 and 1893, the provincial government conducted extensive surveys into the feasibility and practicalities of a sewage farm at Banaras. This was against the personal inclination of Sir Auckland Colvin, who felt that 'except in the driest weather' there was no 'valid objection' to running the sewage into the river right before its confluence with the Varana.[49] Elsewhere, the Lieutenant-Governor gave a more detailed explanation of his view:

> The question of the disposal of the sewage of these Indian cities admits, in my judgement, of further inquiry before it is assumed that the river must be abandoned, and the sewage farm established. There seem to me grounds for contending that, with the great amount of water available in India, and the comparatively small amount of sewage disposed of, the risk of poisoning the river would be minimised, even though the existing system of discharging into the river were enlarged.[50]

At the same time, Auckland Colvin saw himself forced to acknowledge the necessity of a sewage farm, as he felt 'the full force of objections raised on the point of public sentiment, and the further objection that our practice may be quoted against us by the Panjab with weight, should Delhi for example improve its drainage'.[51]

The conditions for the establishment of a farm at Banaras, the surveys revealed, were highly conducive: Land was available at cheap prices, and the soil possessed great absorbing powers and drainage, so that there was hardly any danger of waterlogging. Moreover, the majority of tenants and cultivators were of low caste and therefore were not expected to have any strong 'sentimental objections' against the use of sewage.[52] In December 1893, the sanitary commissioner of the provinces, G. Hutcheson, communicated his conclusion that 'the scheme was perfectly feasible', and could be carried out 'without difficulty and at reasonable cost'. This method of disposing the sewage of Banaras, he added, was not only in line with the principles of sanitary science, but also met 'the wishes and sentiments of the Hindu community who have subscribed largely to prevent the pollution of the Ganges'.[53] Optimism about sewage farming was further enhanced by the experience with sullage farms in Madras, which were being operated successfully since 1869.[54]

Meanwhile, the financing of both the waterworks and the sewerage scheme stood on shaky grounds. By early 1894, the municipal board found itself in a precarious situation, unable to start the repayment of the government loan for the waterworks (the repayment of the sewerage loan was to start in 1900). On the one hand, the money gained from increased and additional taxation had worked out to be insufficient, as had the normal surplus of municipal revenues. On the other hand, the municipality had failed to realise the large private contributions initially envisaged by the Kashi Ganga Prasadini Sabha.[55]

As anticipated by Bireshwar Mittra, the cooperation between the Sabha and the municipal administration had proved difficult. During the early planning-stage already, prominent members of the Sabha informed the board that their contributions were reserved for the actual construction of sewers, and may not be used for covering preliminary expenses, such as surveys.[56] In 1889, things turned more complicated. According to the municipal board's chairman, the nephew and successor to the late Maharaja of Banaras had dissolved the Sabha in 1889 under the influence of Raja Shivaprasad, who had convinced the former that the Sabha's contribution was no longer needed in view of the municipality's own efforts. Consequently, even though the Sabha was revived later that year, numerous subscribers withdrew their offers and spent the money otherwise, and by February 1891, only the late Maharaja's subscription of 50,000 rupees had been realised.[57] There was another important reason for the Sabha's troubles in raising contributions. As the District Collector G. Adams pointed out to the secretary of his provinces, the Sabha had 'clearly laid down from the beginning that its *raison d'être* is the protection of the Ganges from pollution within the limits of the Holy City, and on one occasion it specified them distinctly, though it was hardly necessary, as from the Assi to the Barna Sangam [the confluence of the Varana and the Ganges]'. Therefore it was of no surprise that the majority of its subscribers were in no way ready to pay, seeing that the municipality planned to discharge the sewage above the Varana. Of this point, Adams wrote, the administration had completely lost sight, and now that it pressurised the Sabha into handing out the collected funds, the latter naturally brought this point forward.[58] Thus, despite renewed efforts at gaining subscribers, the Sabha by 1893 had collected only 122,950 rupees, which fell drastically short of the ten lakhs the municipality needed to abolish the new residence tax.[59]

To relieve itself from financial pressure, the Municipal Board called upon the Government of India to extend the period of repayment from 30 to 60 years. In this, the Board was strongly backed by the new Lieutenant-Governor of the North-Western Provinces, C.H.T. Crosthwaite. The city's potential as a centre for cholera dissemination, and 'the manner in which the water-works and sewerage schemes [had] been forced on the Committee' in Crosthwaite's view fully justified the board's request.[60] At the same time, the provincial government declared itself incapable to help the municipality out sufficiently from its own revenues.[61] In response, the Government of India refused to give any concession other than reducing the rate of repayment up to the year 1900, and suggested that the board either increase taxation further or stop the sewerage scheme altogether.[62] Neither of these proposals was acceptable to the Lieutenant-Governor and the municipal board, as public discontent about tax pressure was already high, and the construction of sewers had already begun, with considerable sums of money spent on the latter.[63] Under these circumstances, the only solution open to the municipality was to reduce the costs of the sewerage scheme in whatever way possible. Unsurprisingly, the sewage farm was among the first things to be discarded. As pointed out by sanitary engineer Wilson (A.J. Hughes' successor), the cost for the construction of pipes carrying the city sewage over the Varana – a prerequisite for the construction of a sewage farm – itself amounted to 60,000 rupees. This part of the project, he maintained, had been added solely in order to placate the Kashi Ganga Prasadini Sabha and certain members of the municipal board, and was not necessary for the efficient working of the sewerage system as such.[64] Wilson's proposition raised the indignant protest of the board's new British chairman, who stressed that the discharge of the sewage beyond the sacred precincts of the city had long been insisted on by the subscribers of the Sabha and the municipal board. Moreover, the junction of the Varana and the Ganges itself was held sacred by many, and bathing festivals were routinely held not far from the proposed sewage outfall.[65] The financial argument, however, weighed stronger. In late 1896, the North-Western Provinces approved of the sanitary engineer's proposals, informing him that 'in preparing and forwarding projects in future the necessity of postponing works that [are] not absolutely required should be borne in mind'.[66] In December 1896 finally, it was clearly noted that the sewage farm would 'not be taken up for some years to come'.[67]

Thus, colonial fiscal conservatism acted as the major guiding principle for the choice of sewerage technology in Banaras, preventing the inclusion of a sewage farm as desired by the municipal board and others. That said, a closer look at the board's proceedings and the roles played by the Kashi Ganga Prasadini Sabha and other Banaras citizens reveals that Hindu perspectives on the discharge of untreated sewage into the Ganges were in no way homogenous. For some Hindu members of the board, such as Bireshwar Mittra, sanitary considerations generally seem to have carried more importance than religious ones. The Kashi Ganga Prasadini Sabha, on the other hand, was initially preoccupied with the 'Ganga of Kashi', and only at a later stage called for a sewage farm that would keep the Ganges as a whole free from sewage. The citizens of Banaras, finally, were primarily concerned about the financial hardship under which the water supply and sewerage projects put them, and popular expressions of discontent against these projects were part of a larger framework of protest against the provincial government's ambitious agenda of urban sanitary reform. Nevertheless, the proceedings of the town hall meeting of 27 April 1889, and British officials' repeated allusions towards 'public sentiment' with regard to sewage disposal into the Ganges, show that the issue was definitely of religious concern to some part of the Hindu population. In the end, however, religious arguments carried little weight, and failed to make a lasting impact on finance-driven colonial policy.

The events around the construction of a sewage farm at Kanpur followed a pattern similar to that in Banaras. Writing on the Kanpur sewerage project in 1892, G. Hutcheson and A.J. Hughes maintained that sewage treatment was an integral part of the overall scheme. They acknowledged that the discharge of untreated sewage into rivers could be resorted to only as a temporary measure, one 'made compulsory by the pressure of financial necessity'. This compulsion, however, was very much in force at Kanpur in their view. If the Kanpur municipal board included a sewage farm right away, they warned, it would incur an additional expenditure of 300,000 rupees, plus another annual 18,000 rupees to be spent on interest and the sinking fund. There was no way that the board, whose budget was already strained by the basic water supply and sewerage schemes, could reach an annual income sufficient to cover these additional costs. Thus, Hutcheson and Hughes concluded that '[u]ntil our sources consolidate and plans of taxation develope [sic] there is no other means for disposing of the sewage open to us but putting the sewage into the Ganges'. In preparation, all the works would be designed in a way suitable to the ultimate disposal of the sewage on land, which, they reiterated, 'must be held to be an important and essential feature of the project only postponed for a time'.[68] The Government of India's committee on sewage disposal in 1893 further strengthened the case for a sewage farm. According to the committee, sewage treatment had to be an integral part of all sewerage schemes in India, a sewage farm being the most desirable treatment method. In Kanpur therefore, a sewage farm was to be built to the east of the cantonment, where there was suitable, unsaturated land.[69]

For several years, Kanpur's municipal board sought to remodel its present taxation scheme in order to enhance municipal revenue.[70] Ultimately, however, it found itself at heads with the provincial and central governments over funding, just as the Banaras municipal board did at around the same time. Writing to the Government of India, the Kanpur municipal board claimed that the proposed sewerage scheme could not be executed without a considerable contribution from government, and reminded the latter of its earlier promise not to allow municipal taxation to exceed a certain limit. The Government of India in turn raised several issues in defence, e.g. the fact that the municipality had in fact doubled its tax income. Confronted with this impasse, the Lieutenant-Governor of the North-Western Provinces put the necessity of a sewage farm, the costs of which were estimated to exceed two lakhs rupees, into question. To add a farm, the Lieutenant-Governor maintained, would 'remove any chance that formerly existed of the resources of the municipality being equal to the undertaking [of a sewerage system]'. Financial considerations were further supported by the North-Western Provinces' prevailing ideology on sewage disposal into Indian rivers. According to the sanitary commissioner, the discharge of crude sewage into the Ganges at Kanpur was unlikely to cause any harm due to the great self-purificatory powers of the river, and the absence of towns for many miles downstream.[71] With no financial support forthcoming, the Kanpur municipal board decided to discard not only the sewage farm, but the sewerage scheme as a whole.[72]

Notes

1 Tinker 1968: 28–37, 43–8.
2 Harrison 2011: 290–1.
3 See, for example: Harrison 1980; Mann 2007; Ramanna 2002: 83–122.
4 Ramasubban 1982: 41; Arnold 1985: 167–83.
5 Tinker 1968: 73.

6 Harrison 1994: 200.
7 Harrison 1994; Arnold 1993; Muraleedharan and Veeraraghavan 1995; Basu 1995; Ramanna 1996; Prashad 2001; Mann 2007; see also Khalid (2012: 58–9).
8 Prashad 2001: 155.
9 On the military's heavy impact on colonial finance see Douglas M. Peers' article 'State, Power, and Colonialism' (Peers 2012).
10 Mann 2015a: 296–7, 301–2; Gandy 2008: 113; Harrison 1994: 172, 176–7.
11 Harrison 1994: 166–7.
12 Mann 2007: 22.
13 Prashad 2001: 155.
14 Harrison 1994: 167.
15 Freitag 2010b: 215–20. See also Heitler 1972. Moreover, not only actual protest could exert considerable influence over official public health policy, but also the perceived threat of such a protest in the minds of colonial officials (Khalid 2012).
16 'Proceedings of a public meeting held to concert measures to prevent the pollution of the River Ganges by the discharge of filth into it through the sewers', in *A Collection of Miscellaneous Papers Relating to the Benares Water-Works and Drainage Projects*, p. 2, UPSA, GoNWP, Municipal Dpt, File 3/63/1890, Box 532, 'Benares Water-Supply and Drainage Project'.
17 'Proceedings of a public meeting', in *A Collection*, pp. 2–3.
18 *Bharat Jiwan*, 22 November 1886, p. 3; APAC, IOR, L/R/5/63, *North-Western Provinces Newspaper Reports [henceforth: NWPNR], 1886*, 'Sarosh-I-Benares', 1 December 1886, and 'Akbhar-i-Chunar', 7 December 1886.
19 *ARSC NWP for the year 1886*, p. 65; 'Reply of His Excellency the Viceroy to the address of the Kasi Ganga Prasadini Sabha, December 1890', UPSA, GoNWP, Municipal Dpt., Box 114, File 979A, 'Application from the members of the Kasi Ganga Prasadhini Sabha, Benares, for His Honor's patronage'.
20 B. Mittra, 'Memorandum on the Benares water-supply and sewerage scheme', 16.8.1888, in *A Collection*, p. 3.
21 Joshi 1965: 281–2.
22 Freitag 2010a: 5–16. On the merchant-bankers see also Bayly (1983: 177–80).
23 B. Mittra, 'Memorandum on the Benares water-supply and sewerage scheme', 16.8.1888, in *A Collection*, p. 3.
24 Secy GoNWP to Commissioner, Benares Division, 4.10.1888, UPSA, GoNWP, Municipal Dpt., File 3/63/1890, Box No. 532, 'Benares Water-Supply and Drainage Project', Prgs February 1889, No. 4; 'Extract from the proceedings of a special meeting of the Municipal Board, Benares', 4.1.1889, in *A Collection*, pp. 14–15.
25 'Extract from the proceedings of a special meeting of the Municipal Board, Benares, held on the 4th January 1889', in *A Collection*, p. 15; note by Jas. White, Chairman, Banaras Municipal Board, 6.4.1889, in *A Collection*, pp. 38–9.
26 Supplementary note by Bireshwar Mittra, Member, Banaras Municipal Board, 21.2.1889, in *A Collection*, pp. 24–30.
27 Freitag 2010b: 221–2.
28 'Proceedings of a general meeting of the citizens of Benares held within the town hall compound on the evening of Saturday, the 27th April 1889', APAC, IOR, P/3598, NWP, Municipal Dpt., Prgs June 1890, No. 1; *The Pioneer*, 8 May 1889, pp. 585–6.
29 R. Smeaton, Secy GoNWP, to Commissioner, Benares Division, 20.5.1889, APAC, IOR, P/3598, NWP, Municipal Dpt., Prgs June 1890, No. 3.
30 'Statement of objections raised by the citizens of Benares for the enhancement of octroi duty and other new taxes', UPSA, GoNWP, Municipal Dpt., File No. 227A, 'Revision of taxation, Benares Municipality', Prgs November 1890, No. 11.
31 B. Mittra, 'Memorandum on the Benares water-supply and sewerage scheme', 16.8.1888, in *A Collection*, p. 5. Initially, orthodox Hindus in the capital had refused to use the city's piped water supply, holding that the water was rendered ritually

impure by its transmission through iron pipes, and by the (alleged) fact that the pumping engine was greased with animal tallow (Arnold 2013: 8; see also Chakrabarti 2015: 193–7).

32 *The Pioneer*, 1 May 1889, p. 554.

33 A.J. Hughes, Supervising Engineer, Municipal Water-Works, NWP, to Chief Engineer, Public Works Dpt., NWP, 27.2.1890, UPSA, GoNWP, Municipal Dpt., File 3/63/1890, Box 532, 'Benares Water-Supply and Drainage Project', Prgs June 1890, No. 14.

34 Note by J.G. Forbes, Chief Engineer, Irrigation Branch, 19.11.1889, UPSA, GoNWP, Municipal Dpt., File 3B/63/1890, Box 532, 'Benares Water-Supply and Drainage Project'; R. Smeaton, Secy GoNWP, 16.12.1889, in *A Collection*, p. 58.

35 *Bharat Jiwan*, 29 April 1889, p. 4; APAC, IOR, L/R/5/66, *NWPNR, 1889*, 'Hindustan', 2 May 1889; also APAC, IOR, L/R/5/66, *NWPNR, 1889*, 'Rafi-ul-Akhbar', n.d.

36 APAC, IOR, L/R/5/67, *NWPNR, 1890*, 'Rafi-ul-Akbar', 17 March 1890.

37 Freitag 2010b: 220–1; Kumar 2010: 149–50.

38 APAC, IOR, L/R/5/67, *NWPNR, 1890*, 'Hindustani', 8 June 1890. See also APAC, IOR, L/R/5/67, *NWPNR, 1890*, 'Najmu-l-Hind', 8 June 1890, and 'Hindustan', 30 August 1890.

39 J. White, Chairman, Banaras Municipal Board, 'Demi-Official', 16.4.1891, APAC, IOR, L/PJ/6/301, GOI, Public and Judicial Dpt., Judicial and Public Annual Files, File 907.

40 Freitag 2010b: 222–3.

41 For a discussion on Indian attitudes towards Western technology see Arnold (2000: 123–4).

42 See also Tinker 1968: 75; Freitag 2010b: 215–20; Mann 2015a: 302.

43 Jas. White, Chairman, Banaras Municipal Board, to R. Smeaton, Secy GoNWP, 1.4.1890, in *A Collection*, p. 61; 'Proceedings of a meeting of the drainage and water-works committee, held at Benares on the 9th September 1890', in *A Collection*, p. 72.

44 See Chapter 2.

45 T.W. Holderness, Secy GoNWP, to Jas. White, Chairman, Banaras Municipal Board, 20.9.1890, in *A Collection*, pp. 75–6.

46 Jas. White, Commissioner, Banaras Municipal Board, to T.W. Holderness, Secy GoNWP, 16.10.1890, in *A Collection*, p. 78; A.J. Hughes, Supervising Engineer, Municipal Water-Works, NWP, to T.W. Holderness, Secy GoNWP, 16.10.1890, in *A Collection*, p. 79.

47 Jas. White, Commissioner, Banaras Municipal Board, to T.W. Holderness, Secy GoNWP, 16.10.1890, in *A Collection*, pp. 78.

48 Nevill 1909a: 264.

49 Memorandum by Sir Auckland Colvin, 17.9.1892, APAC, IOR, P/4062, GoNWP, Municipal Dpt., Prgs October 1892, No. 68.

50 Colvin 1894: 522.

51 Memorandum by Sir Auckland Colvin, 17.9.1892, APAC, IOR, P/4062, GoNWP, Municipal Dpt., Prgs October 1892, No. 68.

52 J.O. Miller, Director of Land Records and Agriculture, NWP, to Secy GoNWP, 8.8.1892, APAC, IOR, P/4062, GoNWP, Municipal Dpt., Prgs October 1892, No. 67.

53 G. Hutcheson, Sanitary Commissioner NWP, to Commissioner, Banaras Division, 4.12.1893, APAC, IOR, P/4702, GoNWP, Municipal Dpt., Prgs January 1895, No. 31(f).

54 J. Nield Cook, Health Officer, Madras Municipal Commission, 'Report on the sewage farm system in Madras (Town)', 14.4.1891, APAC, IOR, P/4112, GOI, Home Dpt., Sanitary Branch, Prgs November 1892, No. 7. Even though the report refers to a 'sewage farm' it is evident that this was actually a sullage farm, as it is mentioned that the liquid contained 'comparatively little night-soil'.

55 NAI, GOI, Home Dpt., Municipalities, Prgs May 1895, Nos 46–7.

56 B. Mittra, 'Memorandum on the Benares water-supply and sewerage scheme', in *A Collection*, p. 3.
57 Jas. White, Chairman, Banaras Municipal Board, to R. Smeaton, Secy GoNWP, 9.2.1891, in *A Collection*, p. 80; B. Mittra, Member, Banaras Municipal Board, to Colonel Forbes, Chief Engineer and Secy GoNWP, Public Works Dpt., 21.11.1890, in *A Collection*, p. 54.
58 G. Adams, District Collector, to R. Smeaton, Secy GoNWP, 25.2.1891, in *A Collection*, p. 83.
59 NAI, Home Dpt., Municipal Branch, Prgs September (B), No. 19/21.
60 Demi-Official by C.H.T. Crosthwaite to Sir James Westland, 9.1.1895, NAI, Home Dpt., Municipalities, Prgs, May 1895, Nos 46 and 47.
61 P.K. Mittra, GoNWP, 18.7.1894, NAI, Home Dpt., Municipalities, Prgs May 1895, Nos 46–7.
62 J.E. O'Conor, Assistant Secy GOI, Finance and Commerce Dpt., to Secy GoNWP, 20.4.1895, NAI, Home Dpt., Municipalities, Prgs May 1895, No. 46A.
63 W.H.L. Impey, Secy NWP, to Secy GOI, Home Dpt., 19.1.1895, NAI, Home Dpt., Municipalities, Prgs May 1895, No. 158.
64 Resident Engineer, Sewerage and Water-Works, Banaras, to Secy Banaras Municipal Board, 10.11.1894, APAC, IOR, P/4702, GoNWP, Municipal Dpt., Prgs March 1895, No. 19(a).
65 R.H. Brereton, Chairman, Banaras Municipal Board, to Commissioner, Banaras Division, 7.12.1894, APAC, IOR, P/4702, GoNWP, Municipal Dpt., Prgs January 1895, No. 34(a).
66 Secy GoNWP, Municipal Dpt., to Sanitary Engineer NWP, APAC, IOR, P/4908, GoNWP, Municipal Dpt., Prgs September 1896 (B), No. 43.
67 H.G. Boyce, Sanitary Engineer GoNWP, 'Inspection notes on the Benares sewerage and drainage works, 11th to 16th December 1896', APAC, IOR, P/5127, NWP, Municipal Dpt., Prgs July 1897, No. 26(a).
68 G. Hutcheson, Sanitary Commissioner NWP, and A.J. Hughes, Supervising Engineer, Municipal Works, NWP, 'Joint note on the sewerage of the city of Cawnpore, with special reference to the arrangements for passing the sewage through cantonments', n.d. [1892], p. 5, NAI, GOI, Home Dpt., Sanitary Branch, Prgs April 1893, Nos 17–25.
69 See Chapter 1.
70 APAC, IOR, P/4702, GoNWP, Municipal Dpt., Prgs February 1895, Nos 3–12; APAC, IOR, P/4908, GoNWP, Municipal Dpt., Prgs March 1896, Nos 7–42.
71 The sanitary commissioner's opinion has been quoted at length at the end of Chapter 2.
72 Note by J.J.D. Latouche, Chief Secy GoNWP, 13.1.1896, APAC, IOR, P/4908, GoNWP, Municipal Dpt., Prgs February 1896, No. 2(a); J.O. Miller, Secy GoNWP, to Commissioner, Allahabad Division, 4.4.1898, APAC, IOR, P/5419, GOI, Home Dpt., Municipalities, Prgs May 1898, Nos 8–11.

4 Biological sewage treatment in the United Provinces

Trial and failure

The breakthroughs of bacteriological science in the late nineteenth century ushered in entirely new perspectives on river pollution and sewage disposal. About a decade after Robert Koch's success in isolating the *comma bacillus* from a water tank in Calcutta, US scientists decisively proved the role of sewage-polluted water supplies in transmitting pathogens. Simultaneously, a number of scientists in Europe and the US began to understand the ways in which micro-organisms bring about the decomposition of faecal matter, a discovery which paved the way for a range of new biological methods of sewage treatment.

Following the spread of waterborne sewerage systems, the number of typhoid epidemics in Europe and the US rose sharply during the late nineteenth and early twentieth centuries. In 1893, England's Worthing district registered over 1,400 typhoid cases, including 198 fatalities. Four years later, fever struck in Maidstone, a town some 40 kilometres away from London, affecting almost 2,000 people and killing 132.[1] In the US, Butler (Pennsylvania) registered 1,400 cases in 1903, with 111 casualties.[2] The frequent recurrence of epidemics stirred public health authorities into action. After typhoid struck at Lowell in the early 1890s, the Massachusetts Board of Health commissioned its chief biologist William T. Sedgwick with a systematic investigation into the aetiology of typhoid. After several years of painstaking research, Sedgwick and his team successfully proved that the responsible pathogen was spread through sewage-polluted water supplies.[3] In reaction to Sedgwick's research, several federal states passed their first river pollution laws around the turn of the century. The first of these was Pennsylvania, which prohibited the discharge of untreated sewage into the state's waterways and entrusted the authority of enforcement into the hands of its Department of Health. By 1905, Minnesota, New Jersey and Ohio had passed laws of similar strength.[4]

Sedgwick's research into the aetiology of typhoid also generated a renewed interest into sewage treatment both in the US and in Europe. A second incentive for innovation came from the practical difficulties British cities faced with land treatment. Urban population growth and the spread of waterborne sewerage systems were producing an ever growing amount of sewage, and consequently, land at the cities' fringes that could be utilised for treatment became scarce and expensive. For the sewage of every 500 to 1,000 people, one acre of land was required. This meant that the treatment of the sewage of a city like London called for 40,000 acres of land. Berlin, which operated the largest and most successful farms all over Europe, used 23,000 acres of land in 1896, an area bigger than the city itself. As discussed earlier, sewage farming was moreover fraught with many difficulties, especially in Britain, where soil and weather conditions were not very conducive to sewage irrigation at large. Looking for a remedy to their sewage crisis, London, Manchester and other cities turned towards biological sewage treatment, which was being developed in Britain and the US from the 1880s. The groundwork for the new method was provided by the path-breaking research of the French chemists Theophile Schloesing and Achille Müntz in 1877. Building on ideas of Louis Pasteur, Schloesing and Müntz investigated the nitrification of sewage in soils, nitrification being considered the chief measure of sewage purification since well-nitrified sewage did not putrefy. Against the prevalent belief, held by Edward Frankland and others, that nitrification was a process of chemical oxidation, the two Frenchmen were able to demonstrate that nitrification/purification was brought about by micro-organisms present in the soil, and as such was a biological process. Initially, these findings were largely ignored by the British sanitary community; in the US they were however received with enthusiasm. The Lawrence experiment station, a public health laboratory established by the Massachusetts State Board of Health in the late 1870s in reaction to the state's increasing problems with river pollution, became one of the leading research centres on sewage treatment. Starting in 1888, a team of biologists (among them William T. Sedgwick), chemists and engineers built several indoor and outdoor tanks filled with sands, gravels and soils typical of various regions, and conducted detailed biological and chemical investigations into the processes of purification occurring in these tanks. The studies not only confirmed Schloesing's and Müntz's thesis that sewage purification on land was a biological process, they also provided essential guidance for the design of the first bacteriological filters employed in biological sewage treatment.[5]

Bacteriological filters belonged to a first set of biological sewage treatment methods that relied on the agency of aerobic bacteria (i.e. bacteria requiring oxygen for their growth). These methods essentially mimicked the process of sewage purification occurring on land, but translated it onto a smaller area, a tank. Sewage was poured into open tanks filled with artificial beds of sand, gravel, clinker, slate or other materials. On these materials, bacteria and other organisms grew and purified the sewage.[6] As Christopher Hamlin notes, biological sewage treatment represented 'a conceptual revolution of the principles of sewage treatment [...], the replacement of a philosophy that saw sewage purification as the prevention of decomposition with one that tried to facilitate the biological processes that destroy sewage naturally'.[7] One of the first aerobic treatment methods, the 'contact bed', was devised by William Joseph Dibdin, chief chemist for the Metropolitan Board of Works and its successor, the London County Council, from 1882 to 1897. In this position, Dibdin was responsible for the Thames estuary into which the London sewage flowed. Dibdin's contact bed was essentially a tight box which held a variety of filtering media such as coke, coal or stone, and kept the sewage in contact with these over a certain period of time. His method furnished a good quality effluent, even though the beds clogged up occasionally.[8] Simultaneously, researchers at the Lawrence station devised the 'trickling filter', a filter similar to the contact bed in design, but allowing air to circulate. The trickling filter was further refined in Britain, where the coarse materials needed for it were available in greater abundance.[9]

A second set of biological treatment methods, developed in the 1890s and early 1900s, used closed tanks, relying on the action of anaerobic bacteria (i.e. bacteria able to grow without oxygen) for the decomposition of sewage. Among these were the septic tank, invented by the city surveyor of Exeter, Donald Cameron, W. Scott Moncrieff's cultivation filter bed, Arthur Travis's hydrolytic tank, and the German engineer Karl Imhoff's Imhoff tank. These tanks were utilised for primary sewage treatment. The anaerobic bacteria liquefied the solid elements in the sewage, leaving an effluent that could subsequently be treated more easily through different aerobic filters.

The last major innovation in biological sewage treatment was the activated sludge process, devised by the chemists Gilbert Fowler, Edward Ardern and William Lockett from Manchester in 1914. In this process, sewage was poured into closed, continuously aerated tanks, which allowed great numbers of bacteria to grow and to feed on the ammonia and organic matter. In a second step, the air was either turned off or the sewage was pumped into a clarifying tank. The bacteria, settling to the bottom, cleansed the sewage of solids and left a clear effluent. After a while, the bacteria were pumped back into the activated sludge tank to start their action anew.[10]

In Britain, early biological sewage treatment was received controversially. Sewage treatment at the time was a fiercely competitive field, wherein advocates of land treatment (and its different variations, such as broad irrigation and intermittent filtration) and chemical precipitation were all vying for the favour of cities, which in turn were under considerable legal and political pressure to treat their sewage and safeguard their rivers and water supplies. The irrigationist lobby in particular was hostile towards the new methods, wary to lose the support of the Local Government Board, which for decades had backed land treatment ideologically and financially, and declared it to be the only efficient method to purify sewage.[11] Ultimately, however, cities like London and Manchester managed to pressurise the British government into revisiting the question of sewage treatment. In 1898, the third major commission investigating the problem of sewage disposal in Britain was established, the Royal Commission on Sewage Disposal. The Commission sat from 1898 to 1915 and produced nine reports on different methods and questions of sewage treatment. Thus, when biological sewage treatment reached British India at the turn of the twentieth century, it was very much of an evolving technology, its various methods still being under evaluation.

Out of the Royal Commission's nine reports, four dealt mainly with the purification of municipal sewage, two with the discharge of sewage into tidal waters, and three with the discharge of manufacturing effluents.[12] The Commission's work was based on a thoroughly scientific approach and presented a major breakthrough, with many of its conclusions remaining authoritative until today. Its interim report of 1901, while still inconclusive on many issues, ended the rigid insistence on the superiority of land treatment and gave due credit to biological and other methods of sewage treatment. Its fifth report on the treatment of municipal sewage (1908) lay down guidelines for the design of different sewage treatment units, such as percolating filters and septic tanks, and its eighth report (1912) established the effluent standards which are upheld till today.[13]

In British India, colonial officials welcomed biological sewage treatment with enthusiasm. Like waterborne sewerage systems, biological sewage treatment was perceived as more sanitary and 'modern' than hand removal and trenching, and held out the promise to rid municipalities of their dependency on sweepers. Moreover, it was seen as a cheap alternative to waterborne sewerage systems, the construction of which involved great expenditure for large sewer networks and sewage farms.[14] Finally, the Government of India's official guidelines formulated by the Kanpur committee on sewage disposal in 1893 clearly directed municipalities to treat their sewage, even though the North-Western Provinces had circumvented their implementation so far. Accordingly, several colonial officials in the North-Western Provinces and Bengal started experimenting with different biological treatment methods at the turn of the twentieth century. In Calcutta, official enthusiasm for the new methods was shared by the European mill managers, who installed septic tanks for their work forces in rapid succession. This and the next chapter shed light on how biological sewage treatment was implemented in various riparian cities, and investigate its impact on river pollution and the direction of official river pollution policy.

The earliest experiments with biological sewage treatment in British India were conducted by sanitary engineer Charles Carkeet James in Bombay. In 1895, Carkeet James constructed a septic tank (which he called 'liquefying tank') similar to Donald Cameron's tank at Exeter, but independently from the latter. His experimentation ground was the Leper Asylum at Matunga, which, due to its geographical location, was not connected with the city's main sewerage system. Until then, the asylum's sewage had been discharged untreated on adjacent land where it was used for irrigation. However, this was soon found objectionable as the land got coated with organic matter and the fodder crops grown on it were spurned by the cattle. By subjecting the sewage to septic tank treatment, Carkeet James was able to obtain a liquid good enough for irrigation, and managed to grow different kinds of crops with reasonable economic success. Apart from his liquefying tank, Carkeet James experimented with a variety of combinations between different aerobic tanks and filters, including macerating tanks and contact beds.[15] A small number of experiments with bacteriological sewage and sullage treatment were also made in other provinces from around the turn of century, e.g. with septic tanks in the Central Provinces and Dibdin filters in the Madras Presidency.[16]

In Banaras, the construction of sewers progressed slowly, but by 1904, the main sewer and all the most important lateral sewers were completed.[17] In 1898, the resident sanitary engineer, H. Lane Brown, built a small aerobic contact filter on the lines of Dibdin's contact bed to purify the sullage of two *muhallās* (wards) entering the Ganges above the intake of the waterworks at Bhadaini. In 1904, Brown further set up a combined latrine and experimental purification plant at Chauka Ghat, a district which could not be connected to the main sewerage network. In this, a latrine with 48 seats was built over a liquefying chamber. The liquefied effluent was then to be treated between a continuous percolating filter and a two-chamber contact filter. Reporting on the installation, Brown expressed his hopes that the plant may be useful in furnishing information on the biological treatment of sewage in the 'tropics', especially as the climate of Banaras was marked by both very cold and hot seasons, characteristic for different zones of the subcontinent.[18] Apparently, the purification plant turned out to be a reasonable success, since the sanitary commissioner of the United Provinces in 1909 reported that it 'continue[d] to be working satisfactorily'.[19] However, the more widespread use of bacteriological filters for sewage treatment was deemed unnecessary. Banaras, Brown explained, already possessed an extensive functioning sewerage network, and Baldwin Latham had clearly stated that the city sewage could be disposed of untreated into the Ganges without harm.[20] Consequently, Banaras kept discharging its sewage untreated into the river for many years to come.[21]

In Kanpur, the question of biological sewage treatment presented itself more forcefully. Other than in Banaras, where the municipal board was able to construct the main body of sewers despite severe financial shortages, the Kanpur municipal board had decided to entirely scrap its sewerage project for lack of funds in 1897.[22] Two years later, the matter was taken up again. Evidently, the municipal board stood under considerable pressure to improve the overall sanitary conditions of the city. In 1895, Kanpur industrialists represented through the Upper India Chamber of Commerce urged the United Provinces government for a committee of inquiry to look at the sanitation question, 'with the view of devising a means for the protection of the commercial and manufacturing interests of the city'. The recent outbreak of cholera, they maintained, was clearly traceable to the 'disgraceful' sanitary state of things.[23] The latter also directly affected the interests of the military. As Kanpur was home to one of the major military stations of British India, the municipality's sanitary policy kept being closely watched by the Government of India's military department and the cantonment authorities.[24]

In 1899, the Kanpur municipal board called the sanitary engineer and the sanitary commissioner of the provinces for a survey. In their report, the officials warned that in the present situation, a serious outbreak of disease was imminent. Hence, they urged immediate improvements in the system of removing street sweepings and night-soil, the remodelling and extension of the surface drainage system, and the prevention of river pollution above the bathing *ghāts*. The 'offensive and dangerous condition of the river', they insisted, had not received due attention so far. Numerous drains from tanneries and municipal drains discharged directly into the Ganges above some of the principal bathing *ghāts*, which were frequented daily by thousands of people.[25] The recent shift of the river away from the city had exacerbated the problem, causing the formation of a 'very offensive' backwater at the mouth of the Sisamau sewer below the leather factory Cooper, Allen & Co. and the North-Western Tannery.[26] Thus, instead of merging into the main stream, the foul liquid from the drains passed right along the river banks, seriously affecting the religious ceremonies of the people:

> [A]s a preliminary part of that ceremony it is incumbent upon each to drink a mouthful of the water. What this apparently harmless and simple cere-mony means when carried out at the Sarsaya Ghat at Cawnpore cannot, we should think, be imagined by those who have not to perform the ceremony themselves.[27]

As a remedy, they suggested the construction of a special sewer as part of the main sewerage scheme, which would intercept the wastewater from the factories and from a number of hamlets and convey it into the main sewer. While Baldwin Latham's original plan of the early 1890s had proposed to carry this main sewer underneath through the military cantonment (which, as it will be remembered, the cantonment authorities strongly opposed), the officials now highlighted the potential of biological sewage treatment to solve Kanpur's sewage problem. After being submitted to biological treatment, they maintained, the purified liquid may be discharged into the Ganges above the cantonment without any problems.[28] An earlier report by the same authors strongly commended this 'simple, efficient and economical method' of sewage purification, suggesting the installation of either septic tanks or Dibdin filters. However, they acknowledged that the suitability of both these systems for the treatment of sewage under 'trop-ical' conditions had still to be verified by actual experience. On the one hand, the higher temperatures might positively enhance the reproduction and activity of the bacteria; on the other, excreta in India were staler than in Britain when brought to the tanks, as they were collected in pail depots beforehand. In view of the promising prospects of biological sewage treatment, they proposed the con-struction of experimental tanks and filter beds.[29]

Biological sewage treatment at Kanpur was also strongly recommended by the English sanitary engineer William Santo Crimp. Crimp was a leading figure in sanitary engineering in Britain and had gained considerable experience in dif-ferent methods of sewage treatment over the years. Among others, he had suc-cessfully remodelled the Wimbledon sewage farm, experimented with various ways to treat and dispose of sludge, and in the early 1890s, as a district engineer to the London County Council, had introduced radical changes to the working of the precipitation tanks at the Barking sewage outfall.[30] As early as 1880, Crimp started experimenting with bacteriological filters and since then had designed and built several different devices for biological sewage treatment.[31] In summer 1899, Crimp was invited to India by the municipality of Bombay to report on the drainage and the general sanitary state of the city.[32] Taking advantage of Crimp's presence in India, a number of municipalities, including those of Simla, Surat, Puna and Kanpur obtained his advice.[33] Reporting on Kanpur after his three-day visit in December 1899, Crimp proposed the construction of a main sewer, to be connected to numerous branch sewers and taking care of the sewage from pail depots, latrines and urinals, and also the sullage from surface drains. Due to the municipality's financial constraints, Crimp thought it most desirable that a gravitation system be adopted, to save the costs for pumping machinery. This, however, limited the choice for the location of the main sewer outfall to some point above the southern boundary of the cantonment. As per the treatment of sewage, Crimp recommended the installation of a macerating tank to deal with the copious amounts of sand contained in Indian sewage (Indians widely used sand for cleansing utensils). This preliminary treatment, according to Crimp, was sufficient in the case of Kanpur. Were the sewage to be discharged into a very small stream used for domestic purposes, it would be necessary to treat it further by filtration or on land, Crimp explained. But as the water volume of the Ganges at Kanpur was large, the final purification was effected rapidly and efficiently by the river itself.[34]

Crimp's choice for the location of the main sewer outfall did not find the favour of the sanitary commissioner and the sanitary engineer. There was no low-lying land close to the cantonment boundary where purifying tanks could be constructed, they argued. Moreover, only about one mile below the site there were a number of important bathing *ghāṭs*, and a mere preliminary purification of the sewage through macerating tanks was therefore insufficient to protect the health of bathers. Also, to carry the sewage underneath through the cantonment would involve an extra expenditure of over two lakhs, which would put many other important sanitary works in abeyance for years. What was needed, according to them, was a low-lying site close to the river, reasonably far away from the city as well as the bathings *ghāṭs*, the cantonment and other settlements. The plot of land between the canal and the railway, they believed, fulfilled all these conditions and was therefore ideally suited for the purpose.[35]

Following the advice of the provincial sanitary officers and William Santo Crimp's, the Kanpur municipal board commissioned its resident sanitary engineer W. Parry with the construction of an experimental purification plant. This included two macerating tanks built along Crimp's design for the preliminary treatment of the sewage, and two filter tanks for secondary treatment. In June 1900, the experiments were commenced with sullage coming in from the Sisamau sewer. Operations proved difficult however. Chemical analyses taken from effluents after purification showed unsatisfactory results and it was decided to add another filter, and to connect one of the macerating filters to a septic tank. Yet, results were still not promising. The reason for this, according to W. Parry, was the character of the sullage, which contained not only night-soil and urine, but also tannery and slaughterhouse wastes, including the blood of slaughtered animals, and cesspool washings. In an attempt to deal specifically with tannery effluents, an experimental settling tank using cinders and brick ballast as filtering media was built at the North-West Tannery. According to Parry, the tank worked well for about three months, purifying the liquid 'from a very dark deep red colour to that of a colour like light Pilsener beer', but the filters soon became clogged. As it was found impossible to treat the tannery refuse with only tank, experiments were discontinued.[36]

With the failure of biological sewage treatment even on a small experimental scale, it appeared unlikely that the whole of Kanpur's sewage could be purified thoroughly enough to justify its discharge into the Ganges above the cantonment, as initially proposed by the sanitary commissioner and the sanitary engineer in their report. Confronted with this failure, as well as restricted funds, the Kanpur municipal board decided to revert to a simplified version of Baldwin Latham's original plan: the construction of a main sewer, intercepting the sullage of the Sisamau drain, numerous other city drains, as well as the factory drains, and, running underneath the cantonment, discharging its contents into the Ganges below cantonment limits. For the time being, excreta would still be collected manually, and no branch sewers, pail depots, latrine and house connections were planned for. A sewage farm, it was pointed out, was way beyond the municipality's budget. Since the liquid to be discharged was thus sullage containing no excreta (apart, of course, from the not inconsiderable amounts finding their way into the drains some way or other), the sanitary commissioner as well as the Lieutenant-Governor of the United Provinces, Sir MacDonnell, were confident that its discharge at the chosen site below the cantonment posed no risk to the health of the people. In early 1902, the Kanpur municipal board took up constructions.[37]

While experimenting with biological sewage treatment, the treatment of sullage mixed with factory effluents, and particularly effluents from tanneries, had come to the fore as a specially difficult undertaking. In his report on Kanpur, William Santo Crimp described tannery refuse as 'a liquid which is of the most objectionable character in consequence of the admixture of lime and other chemicals with refuse highly charged with organic matter of animal origin'. The treatment of this kind of wastewater, he maintained, was extremely difficult. The only option left to the Kanpur municipality, Crimp suggested, would be to precipitate all solid matter in the sullage in settling tanks, and discharge the remaining liquid far into the main stream of the Ganges through a temporary structure, to bring about a high level of dilution.[38] Awareness about the problematic of sullage treatment was further heightened in connection with a new drainage scheme devised for Lucknow just a few years later.

Lucknow, a city with about 250,000 inhabitants at the turn of the twentieth century, lay at the Gomti, described by one official as a 'sluggish stream' with a very small water volume at certain times of the year (see Figure 4.1). To discharge the city sullage into this stream within city limits was considered dangerous to people's health in and below the city, and authorities therefore looked for an alternative method.[39] Like elsewhere, sullage in Lucknow contained dyes and other trade refuse, and its constitution varied greatly from day to day, and from hour to hour.[40] Initially, the construction of intercepting sewers and a sullage farm was discussed, but the idea was quickly dropped since the costs were found

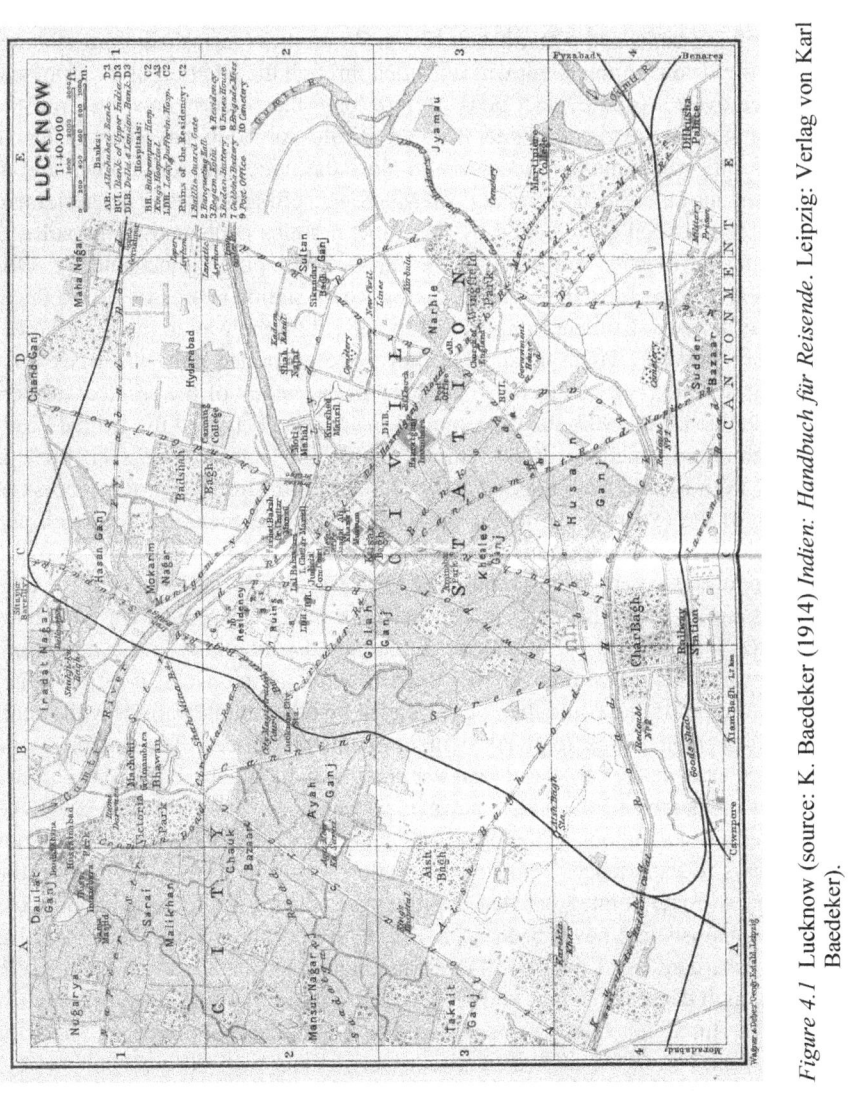

Figure 4.1 Lucknow (source: K. Baedeker (1914) *Indien: Handbuch für Reisende*. Leipzig: Verlag von Karl Baedeker).

prohibitive. As in Kanpur, biological treatment appeared to offer an affordable way out. If successfully applied, filtration devices could be attached right to the outfalls of the existing municipal drains, purifying the sullage on the spot and obviating the necessity of expensive sewers. With this in mind, experiments on the treatment of sullage were taken up in the city in 1905.[41]

Initially, the Lucknow experimental station at Maulviganj was comprised of six liquefying tanks and a number of filters. When these furnished unsatisfactory results, the whole arrangement was extended through the addition of a septic tank and three extra sets of contact beds. Before long, the smell emanating from the septic tank gave rise to complaints from the Riful-i-Am Club and others residents near the site, which made it necessary to close the liquefying tanks down and to turn them into mere detritus tanks.[42,43] Moreover, effluents steadily deteriorated and the contact bed got clogged by the mineral solids contained in the sullage, preventing nitrification and causing the filters to act purely mechanically. Difficulties also arose from the fact that the Lucknow station possessed no chemical analyst of its own. Effluent samples had to be sent all the way to the government laboratory in Agra, which meant that they were examined only one week after they had actually been collected. Therefore the accuracy of the results was seriously doubted.[44] In 1907, the Government of India on behalf of the United Provinces requested sanction for the employment of a 'sewage works chemist or analyst' from Britain from the Secretary of State Lord Morley. For the successful working of the experiments, it was pointed out, the presence of a chemical *cum* bacteriological examiner on the spot was essential. In order to develop systematic knowledge and reach to a standardised process, it was indispensable to take hourly samples for daily analyses, especially in India, where bacterial processes were vitally influenced by alterations in temperature, humidity and evaporation. The 'sewage chemist' moreover was to train others in this work, creating a first set of skilled staff that could then be employed by other municipalities with drainage or sewerage works to oversee their treatment devices. The person to be employed was desired to be of 'similar qualifications to the chemists who are engaged on nearly all the larger sewage purification works at home, and who make daily examinations of the processes employed', and to have a minimum of three years of experience.[45] However, sanction was denied. In his response, the Secretary of State pointed out that the proposed scheme for training subordinate sewage chemists had never been approved of. As per the main sewage chemist, Morley was convinced that a suitable candidate could be found in India itself, since no high degree of expert knowledge was required.[46] Thus, a sewage chemist was never hired for Lucknow, and samples continued to be sent to Agra for examination.[47] In 1909, after several additional trials with different machinery, the failure of the experimental biological treatment plant was acknowledged, and the disposal of the sullage on land was held out to be Lucknow's only feasible alternative.[48] Consequently, the municipality built two intercepting sewers after all, one of them discharging the sullage on waste land, the other on cultivated lands.[49]

Biological sewage treatment, received with great enthusiasm in the United Provinces at the turn of the twentieth century as a cheap alternative to expensive sewer networks and sewage farms, did not solve the municipalities' sewage and sullage disposal problems. While Banaras succeeded in applying biological methods on a small scale, both Kanpur and Lucknow had to abandon them after experiments on a larger scale furnished only disappointing results. The reasons for this failure for one were certainly technical. The treatment of industrial effluents for example, or sullage containing such effluents, presented as much of a challenge to authorities in Europe as to those in India, and at the time was being extensively investigated by the Royal Commission on Sewage Disposal in Britain.[50] However, the success of the Lucknow experimental station and the rearing of a specialised staff of 'sewage specialists' to oversee municipal sewage and sullage treatment in the United Provinces were also considerably curbed by the colonial state's fiscal conservatism. After the Secretary of State's refusal to sanction the hiring of a 'sewage specialist' from Britain, neither the Government of India nor that of the United Provinces showed themselves ready to arrange a suitable candidate from India, which is evident from the fact that the issue was shelved after the Secretary of State's negative reply. Without prospects for an adequately trained staff, the future of sewage and sullage treatment in the United Provinces looked blank.

Confronted with the difficulties of biological sewage treatment, the United Provinces in 1908 invited the renowned British sanitary engineer Gilbert J. Fowler to visit the Lucknow experimental station. In 1899, Fowler had been appointed superintendent and chemist of the Manchester Corporation's sewage works, after working as a chemist and bacteriological assistant for the city council's rivers committee for three years. During his occupation at Manchester, which was to last 20 years, he conducted extensive studies into different methods of sewage treatment.[51] Together with the chemists Edward Ardern and William Lockett, Fowler in 1914 devised the activated sludge process, the last major innovation in biological sewage treatment.[52] Moreover, during a previous visit to India, he had reported on septic tank installations in Calcutta.[53] His task at Lucknow was twofold: For one, he was to suggest affordable technical improvements that would ensure the success of the Lucknow experimental station. Second, Fowler was requested to give his opinion on the fundamentals of sewage and sullage disposal in India, the different points being: (i) The best way to dispose of and treat sullage and sewage; (ii) Whether biological sewage treatment by way of percolation filters or contact beds was advisable for India; (iii) How to free sewage from harmful microbes before discharging it into rivers; and (iv) The disposal of sullage on land to irrigate crops. Fowler's conclusions make it clear that he did not view biological sewage treatment as the most suitable method for India. On the one hand, the Lucknow experiments had plainly revealed the difficulties involved with biological sewage treatment. On the other hand, Fowler pointed out that the introduction of systems of sewerage and sewage disposal lay beyond the financial scope of most Indian municipalities. He therefore recommended that the common practice of trenching night-soil should be continued, and that sullage be disposed of through open drains on sullage farms. As per the current discharge of sewage and sullage into rivers, Fowler noted:

The admixture of the polluting liquid with the river water and the rate of flow at the point of discharge should be such [...] that sufficient dissolved oxygen is present inoffensively to dispose of all objectionable matter, and [...] that no permanent deposit of solid matter can take place.

Thus, dilution was to take care of pollution. Where rivers carried too little water, Fowler recommended a preliminary treatment of sullage (or sewage, respectively) either on land or by one of the biological methods, and the additional sterilisation of the effluent with chloride of lime.[54]

The failure of the Lucknow experimental station and Gilbert Fowler's report reduced biological sewage and sullage treatment in the United Provinces to an auxiliary method at best. Fowler's directions moreover essentially reconfirmed the basic approach adopted by the United Provinces many years before: Sewage disposal into rivers was unproblematic as long as sufficient dilution could be ascertained, i.e. in places like Banaras and Kanpur, but had to be carefully considered in places like Lucknow, where rivers carried little water. With regard to the latter cases, the initial hopes put into biological sewage treatment made way for a renewed enthusiasm for sullage and sewage farms. In 1908, the United Provinces sanitary conference at Nainital strongly commended the utilisation of sullage for the irrigation of crops,[55] and Lucknow's two sullage farms were followed by a sullage farm in Allahabad a couple of years later.[56] As the concluding chapter of this book is going to show, these treatment facilities were to present a challenge of their own.

Notes

1 Benidickson 2007: 156–7.
2 Tarr 1996: 123.
3 Melosi 2008: 92.
4 Tarr 1996: 122–3, 160–1. Existing sewerage infrastructures were exempted from these provisions; however, they applied to extensions made to existing sewers (ibid.).
5 Schneider 2011: 7–16. While American sanitarians and engineers had drawn heavily on British investigations, debates and technological innovations in sewage treatment during most of the nineteenth century, the Lawrence station initiated the two-way circulation of knowledge between the US and Britain (Tarr 1996: 185; Melosi 2008: 109–10).
6 Schneider 2011: xxvi.
7 Hamlin 1988b: 190.
8 Hamlin 1988b; Melosi 2008: 109.
9 Melosi 2008: 109.
10 Schneider 2011: xxvi.
11 Ibid.: 15–16; Hamlin 1988c.
12 Royal Commission on Sewage Disposal 1915, NAI, GOI, Home Dpt., Sanitary Branch, Prgs April 1893, Nos 17–25.
13 Sidwick and Murray 1976; Benidickson 2007: 218–22.
14 See, for example: *ARSC NWP for the year 1891*, p. 10A; G. Hutcheson, Sanitary Commissioner NWP and A.J. Hughes, Supervising Engineer, Municipal Works, NWP, 'Joint note on the sewerage of the city of Cawnpore, with special reference to the arrangements for passing the sewage through cantonments', n.d. [1892], pp. 11–12; *Annual Report of the Sanitary Commissioner with the Government of India*

[henceforth: ARSC GOI] for the year 1900, p. 133; *ARSC Bengal for the year 1904*, pp. 27–8; Clemesha 1910: 5–7; and the following chapters. Sweepers were crucial for the functioning of the urban sanitation regime where no waterborne systems existed, and by going on strike they could put considerable pressure on the colonial administration (Khalid 2012; Masselos 1982).

15 Carkeet James 1906a: 154–98. The macerating tank was designed by the English sanitary engineer William Santo Crimp and worked on similar lines like a septic tank. However, Carkeet James claimed that the macerating tank was chiefly useful in arresting the solids present in sewage, while bringing about only a slight purification of the effluent (ibid.: 170–1).

16 APAC, IOR, P/7059, GOI, Home Dpt., Sanitary Branch, Prgs October 1905, Nos 365–84.

17 H. Lane Brown, 'A brief history […] of the Benares sewerage works, together with some particulars of sewage and sullage purification by bacterial agency as carried out to date', n.d., APAC, IOR, P/7059, GOI, Home Dpt., Sanitary Branch, Prgs October 1905, No. 938. See also Nevill 1909a: 264–5.

18 H. Lane Brown, 'A brief history […]', APAC, IOR, P/7059, GOI, Home Dpt., Sanitary Branch, Prgs October 1905, No. 938.

19 *Annual Report of the Sanitary Commissioner of the United Provinces [henceforth: ARSC UP] for the year 1909*, p. iii. Lane's report suggests however that no systematic inquiries were made to assess actual purification efficiency.

20 H. Lane Brown, 'A brief history […]', APAC, IOR, P/7059, GOI, Home Dpt., Sanitary Branch, Prgs October 1905, No. 938.

21 See this book's conclusion.

22 J.S. Meston, Secy GoUP, to Secy GOI, Home Dpt., 13. and 16.3.1901, APAC, IOR, P/6762, UP, Municipal Dpt., Prgs May 1904, No. 1.

23 W.B. Wishart, Secy Upper India Chamber of Commerce, Kanpur, to Chief Secy GoNWP, 5.9.1895, APAC, IOR, P/4908, NWP, Municipal Dpt., Prgs February 1896, No. 35.

24 See, for example, NAI, GOI, Home Dpt., Sanitary Branch, Prgs April 1893, No. 17–25.; and NAI, GOI, Home Dpt., Municipal Branch, Prgs May 1897, Nos 23–6.

25 Note by W.B. Gordon, Sanitary Engineer NWP, and W.G. Thorold, Offg. Sanitary Commissioner NWP, 19.3.1900, APAC, IOR, P/6297, NWP, Municipal Dpt., Prgs January 1902, No. 2(a).

26 *ARSC NWP for the year 1898*, p. 43B.

27 Note by W.B. Gordon, Sanitary Engineer NWP, and W.G. Thorold, Offg. Sanitary Commissioner NWP, 19.3.1900, APAC, IOR, P/6297, NWP, Municipal Dpt., Prgs January 1902, No. 2(a).

28 Ibid.

29 Joint note by the Sanitary Engineer and Offg. Sanitary Commissioner NWP, 24.8.1899, APAC, IOR, P/5823, NWP, Municipal Dpt., Prgs February 1900, No. 57(a).

30 w.a. 1901a; w.a. 1901b.

31 William Santo Crimp, 'Report on main drainage scheme', n.d., APAC, IOR, P/6297, NWP, Municipal Dpt., Prgs January 1902, No. 3.

32 The metropolis had recently been ravaged by an epidemic of bubonic plague, which precipitated the greatest social and economic crisis in British India since the Great Rebellion. In 1898, the Bombay Improvement Trust was founded with the aim of restoring the 'health' of the city (Kidambi 2007: 49–114).

33 w.a. 1901b: 345–6.

34 William Santo Crimp, 'Report on main drainage scheme' scheme', n.d., APAC, IOR, P/6297, NWP, Municipal Dpt., Prgs January 1902, No. 3.

35 Note by W.B. Gordon, Sanitary Engineer NWP, and W.G. Thorold, Offg. Sanitary Commissioner NWP, 19.3.1900, APAC, IOR, P/6297, NWP, Municipal Dpt., Prgs January 1902, No. 2(a).

36 'Report by W. Parry [...] on the sewage experimental filtration tanks, Cawnpore', APAC, IOR, P/7059, GOI, Home Dpt., Sanitary Branch, Prgs October 1905, Nos 365–84.
37 L.C. Porter, Magistrate and Chairman, Kanpur Municipal Board, to Commissioner, Allahabad Division, 11.1.1901, APAC, IOR, P/6059, NWP, Municipal Dpt., Prgs October 1901, No. 5(a); J. Meston, Secy GoNWP, to Secy GOI, Home Dpt., 13. and 16.3.1901, APAC, IOR, P/6762, UP, Municipal Dpt., Prgs May 1904, No. 1.
38 William Santo Crimp, 'Report on main drainage scheme' scheme', n.d., APAC, IOR, P/6297, NWP, Municipal Dpt., Prgs January 1902, No. 3.
39 A.R. Sutherland, Secy GoUP, to Secy GOI, Public Works Dpt., 9.10.1907, NAI, GOI, Home Dpt., Sanitary Branch, Prgs January 1908, No. 351.
40 *ARSC UP for the year 1906*, pp. iv–v.
41 A.R. Sutherland, Secy GoUP, to Secy GOI, Public Works Dpt., 9.10.1907, NAI, GOI, Home Dpt., Sanitary Branch, Prgs January 1908, No. 351.
42 A detritus tank serves to remove the large heavy suspended matter from sewage (Williams 1924: 60–4).
43 A.W.E. Standley, 'Note on an experiment on sullage treatment at Lucknow' in *The Proceedings of the Second All-India Sanitary Conference held at Madras, November 11th to 16th, 1912, Vol. II, Hygiene*, pp. 428–30.
44 Standley, 'Note on an experiment on sullage treatment at Lucknow'.
45 APAC, IOR, P/7886, GOI, Home Dpt., Sanitary Branch, Prgs January 1908, Nos 351–3.
46 John Morley, Secretary of State, 28.2.1906, APAC, IOR, P/7881, GOI, Home Dpt., Municipalities, Prgs May 1908, No. 29.
47 Standley, 'Note on an experiment on sullage treatment at Lucknow'.
48 Ibid., p. 430.
49 P.R. Hewlett, 'A short descriptive note on Lucknow city sanitation', in *The Proceedings of the Third All-India Sanitary Conference held at Lucknow, January 19th to 27th, 1914, Vol. V, Papers*, p. 73.
50 See, for example, the Royal Commission's ninth report (published in 1915); see also Carkeet James 1906a: 197–8.
51 Watson 1953.
52 Schneider 2011: xxvi.
53 On Fowler's work in Calcutta see Chapter 5.
54 G.J. Fowler, 'Report on the Treatment of Sullage in the United Provinces', APAC, IOR, P/8380, GoUP, Municipal Dpt., Prgs March 1910, No. 11.
55 'Proceedings of the meeting of the Municipal Sub-Committee, 10.9.1908' and J. Chaytor-White, Note on disposal of night-soil, 23.8.1908, APAC, IOR, V/27/840/32, 'United Provinces Sanitary Conference. Collection of papers relating to the Sanitary Conference'.
56 *ARDPH UP for the year 1916*, p. iv.

5 Biological sewage treatment in Calcutta

The septic tank controversy

During the 1890s, debates around river pollution through sewage in British India principally arose from the United Provinces' ambitious agenda for urban sanitary improvement. The advent of biological sewage treatment at the turn of the twentieth century shifted the debate from the United Provinces to Bengal, and more specifically to the capital of the Raj, Calcutta. In the United Provinces, official experiments with biological sewage treatment were short-lived and provoked no noticeable reaction from the Indian public. In Calcutta, biological sewage treatment was adopted extensively by factories along the Hooghly,[1] which built septic tank latrines to dispose of the excreta of their workforce and discharged the resulting effluents directly into the river. This practice revived official debates around the impact of sewage – or, in this case, allegedly treated sewage – on river water quality and public health, and also infuriated parts of the local Indian population. Thus, Calcutta witnessed a most heated debate around sewage disposal and river pollution during the 1900s, which was similar in intensity to the debates of the 1890s, but involved the Indian public at a much larger scale.

This chapter investigates the Bengal government's policy on biological sewage treatment and river pollution, and examines the different factors that shaped it. In doing so, it connects to the analytical structure and the central themes laid out during previous chapters. For one, it analyses the roles and perspectives of British and Indian officials at different administrative levels. As it will be noticed, the administration most heavily involved in this debate was the provincial government of Bengal. Municipal administrations played only a marginal role, as the mills employing septic tank latrines were scattered over several municipalities, and the issue therefore had to be dealt with by a centralised authority. Nevertheless, the chapter zooms into the 'ground-level' of the city, where two principal factions were actively trying to influence the direction of government policy: members of the Indian public and the Indian press, on the one hand, and the European mill managers, on the other. Second, the chapter sheds light on the continuity of two central themes which had emerged during the 1890s, and which both essentially represent different attempts to define 'pollution' and to determine its impacts on the Ganges: debates on water quality and public health, intricately related to questions of disease aetiology and oriented along criteria of contemporary science, and resistance against official policy based on a religious, Hindu understanding of 'pollution'. Thus, with the case study presented here I contend that while the context for debates on sewage disposal and river pollution during the first decade of the twentieth century was markedly different from that prevailing during the early 1890s – based on the far-reaching epistemological shifts with regard to disease theory and sewage treatment during the 1890s and early 1900s, and due to the particular political, social and environmental conditions prevailing at Calcutta – colonial policies showed important continuities.

First, the 'Indian paradigm' kept perpetuating colonial attitudes towards river pollution. Following this paradigm, many British Bengal officials viewed river self-purification and dilution as specially powerful under Indian environmental conditions, and thereby justified the discharge of large volumes of septic tank effluents into the Hooghly, despite their acknowledged problematic bacterial contents. Second, the Bengal government perpetuated colonial attitudes towards Indian resistance against sewage disposal into the Ganges, as evident earlier in Banaras, forcing through its agenda of sanitary and technological 'progress'. Most importantly, it followed the government of the United Provinces in moulding its policy solely along scientific parameters, declining to value Hindu notions of 'pollution' and the resistance that built on them.

From its foundation as a trading settlement by the East India Company in 1690, Calcutta by the twentieth century had developed into a major colonial metropolis, counting roughly 850,000 inhabitants according to the census of 1901 (see Figure 5.1).[2] The city derived its stature both from being the administrational seat to the Government of India and the Bengal province, and from the key role it played for imperial commerce. Situated on the east banks of the Hooghly some 130 kilometres from the sea, the city and its port acted as a nodal point from which the products of Bengal's hinterland – jute, rice, muslin and silk yarn, to name just a few – were exported to the international market.[3] Due to its political and economic importance, its large European population, and its location in the Bengal Delta (the region known as the cradle of cholera epidemics) Calcutta was early put under scrutiny by colonial sanitary reformers. In 1835, surgeon James Ranald Martin conducted a first survey into the city's sanitary condition. His report paved the way for a second, massive investigation by the newly established Fever Hospital and Town Improvement Committee in 1840, which rendered a dismal picture of the present drainage, cleanliness, ventilation and water supply of the city. While the Committee's report raised considerable attention from the public and the administration, it was not followed by any immediate action due to lack of funds.[4] A couple of years later, the Committee's work started to bear fruit. Designed by the chief engineer of the Calcutta Corporation, William Clark, the core of the municipal drainage and sewerage system was constructed between 1860 and 1875. A big intercepting sewer, eight feet high and six feet wide, was laid out underneath the city's principal streets and connected to smaller underground cross sewers. Up to 1896, this system was subsequently modified and supplemented, ultimately covering an area of 19.2 square kilometres of the central town. The Suburban Sewerage Scheme added 32

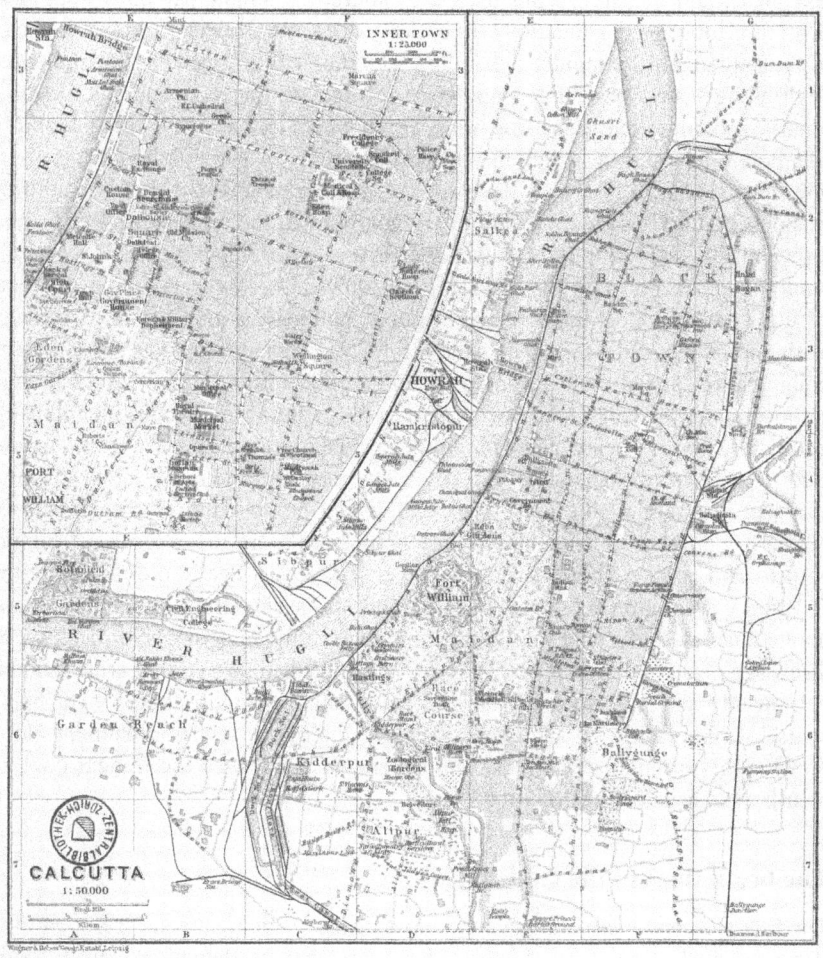

Figure 5.1 Calcutta (source: K. Baedeker (1914) *Indien: Handbuch für Reisende.*
Leipzig: Verlag von Karl Baedeker).

square kilometres between 1891 and 1906, and included the newer southern
areas of the city. The underground network carried the sewage to two pumping
stations at Palmer's Bridge and Baliganj, from where it got pumped into high-
level sewers and discharged by gravity into the Raja *khāl*, a creek of the tidal
river Bidyahari. The working of the sewerage system was intrinsically dependent
on the city's water supply. From 1870 onwards, the waterworks at Palta, some
30 kilometres to the north of the city, distributed water taken from the Hooghly
after submitting it to filtration.[5]

There are no indications that either colonial officials, British or Indian residents objected to the discharge of Calcutta's sewage into the Bidyahari river.

The reason for this most likely is that there were no large settlements along the Bidyahari, and that it was not used as a source for municipal water supplies.[6] With the Hooghly, the situation was entirely different. Flowing through the heart of the capital and its rapidly expanding suburbs, it functioned as a main source of drinking water for a large population. Moreover, the river due to Calcutta's relative proximity to the sea was tidal, which meant that anything dumped into it was bound to be carried back through the interplay of ebb and flow. Therefore, the more sanitary awareness developed, the more keenly the river's condition was observed by the colonial administration.[7] One of the main incentives to start a sewerage system in the 1860s was the Bengal government's desire to stop the dumping of excreta into the river, which hitherto had been common practice. *Mehters* would collect the contents of public and private privies and carry them to the so-called night-soil *ghāṭ* near the Mint. From there, boats hired by the municipality carried the load downstream and put it into the river. Even though authorities suspected that the greater part of the excreta was clandestinely thrown into public drains, the amount officially disposed of into the river was still estimated to account to 200 tons every day.[8] The reorganisation of excreta disposal was just one part of the more wide-ranging effort to improve the Hooghly's sanitary condition. Other major issues of concern were the disposal of human corpses and animal carcasses into the river, a practice legally prohibited by the Calcutta Corporation in 1864, and the pollution issuing from municipal drains, public latrines, factory latrines and people defecating along the river banks. The discharge of sewage from Fort William, which housed up to 3,000 army personnel, raised indignant protests from Bengal's sanitary commissioners over many years.[9] As the following sections will show, biological sewage treatment along the Hooghly through septic tanks fundamentally challenged the Bengal government's established river pollution policy.

The beginnings of the controversy

The first systematic experiment with biological sewage treatment in Bengal was taken up at the initiative of the sanitary engineer of Bengal, Albert E. Silk, during the closing years of the nineteenth century. The need for a new latrine for Indian warders at the Presidency Jail in Calcutta offered a suitable opportunity for experimentation, and latrines connected to septic tanks were installed in the jail's garden. In line with other officials, Silk's enthusiasm for biological sewage treatment stemmed to a great part from its promise to substitute for the 'horrible nuisance and expense' of removing night-soil by carts and the 'horrors' of trenching grounds. Reporting on his experiment, the sanitary engineer highlighted the success of the installations at Presidency Jail, having dealt efficiently with the excreta of around 220 sittings per day and creating no nuisance to the surroundings. Moreover, the installations had been very easy to maintain. Nevertheless, Silk admitted that there was still ample scope for improvement, as the effluent produced by the septic tanks was literally 'swarming with bacteria'.[10] The latter statement was in reference to the bacteriological investigations conducted by L. Rogers, professor of pathology at the Calcutta Medical College. In his report, Rogers stated that the effluents were indeed non-putrescible, but while the numbers of bacteria present in the original sewage, both harmless and pathogenic ones, got markedly reduced by the septic tank process, their decrease was insufficient. Most importantly, pathogenic bacteria like those of the *coli* group were reduced at a slower rate than harmless bacteria. In a further experiment, trials were made to purify the effluent further by applying it to land, but this too brought no sufficient improvement. As a result, Rogers expressed his conviction that the septic tank effluent had to be considered as nothing less than diluted, potentially very dangerous sewage, and should under no circumstances be passed into a water body used for drinking water supply. The negative verdict of the bacteriologist was unanimously supported by the chemical examiner of the provinces, who had also analysed the effluents. Confronted with these results, the inspector-general of hospitals of Bengal appealed to the government to abandon the experiments. In his request, he was supported by the jail authorities, who complained that the large open water tank in the jail garden used for the disposal of the septic tank effluent had been turned into a veritable cesspool, posing a serious threat to public health.[11] According to the inspector-general and other critics, the septic tank installations recently set up at Fort William had similarly revealed numerous problems connected to the working of this still experimental technology.[12] In response, the Bengal government directed the sanitary board of the province (in which A.E. Silk served as secretary) to conduct an inquiry on the spot. At the same time, the Lieutenant-Governor noted that the septic tank system was 'most valuable and useful as compared with the present system of removal and burial of night-soil', provided the effluents were carried off in such a way as not to compromise the water supply of the neighbourhood.[13] Given the promising prospects of sewage treatment through septic tanks, the Lieutenant-Governor further sanctioned the establishment of an experimental tank by the Muzaffarpur municipality, the first municipal venture of this kind in Bengal.[14]

Following government orders, the sanitary board inspected the installation at the Presidency Jail and, satisfied by its observations, resolved that these did not present a nuisance in any way. In contrast, they had shown that sewage disposal by septic tanks had passed the experimental stage and now was a 'proved working success'. As per the quality of the septic tank effluents, the board did not make any further comments, but directed their discharge into the Calcutta sewers through watertight drains, in order to prevent the contamination of water supplies.[15]

Meanwhile, the managers of Calcutta's jute mills had discovered septic tanks as a new convenient method to dispose of the excreta of their work force. During the second half of the nineteenth century, the capital of British India had developed into the hub of the Indian jute industry, and by the early twentieth century figured as the leading jute manufacturing centre in the world. Jute was primarily used as a packing material for industry and trade, especially in the form of gunny bags, and its significance only declined with the discovery of a synthetic substitute in the mid-twentieth century. Other than the Indian-dominated cotton industry at Bombay, the jute industry at Calcutta until World War I was managed by powerful British, mainly Scottish, managing houses. These were large firms not only involved with jute but also with various other trades, managing tea estates, coal mines and steamship companies for inland and overseas trade. The first jute spinning machinery was set up in 1855 at Rishra near Serampore by the British coffee planter George Acland, who took advantage of the increased demand for jute products during the Crimean War caused by the interruption of Russian hemp supplies. By 1875, 16 more mills had followed the footsteps of the Rishra Jute Mill, and by 1911, India possessed a total of 54 mills, all of them located in Calcutta. In order to guard their interests in the jute business, the managing agents in 1884 founded the Indian Jute Mills Association, with the purpose to limit the number of new mills, to prevent recession and price cuts, and to regulate production and prices in order to control competition. In time, the Indian Jute Mills Association became a powerful lobby, represented in the Bengal legislature and well-connected with the colonial administration.[16]

The majority of jute mills were situated along both banks of the Hooghly on strips of land about 60 miles long and 2 miles broad, thus being situated to the north as well as to the south of the city centre (see Figure 5.2).[17]

Apart from jute mills, a number of other factories were placed along the river, among them paper and cotton mills. In 1894, there existed all in all 56 factories, 30 on the Howrah side and 26 on the Calcutta side.[18] These mills concentrated an enormous work force. In 1912, the jute industry alone employed roughly 200,000 workers.[19] Before discovering the septic tank, the Calcutta mills followed the traditional method of conservancy, i.e. excreta were removed manually and brought to trenching sites by sweepers. With a work force often comprising several thousand people, this was a considerable task, and many managers therefore naturally welcomed septic tanks as a cheap and convenient alternative. Thus, while government experimentation with the new technology was still in progress, a fast-growing number of mills were installing septic tanks on their own accord. In December 1902, the sanitary engineer together with the sanitary commissioner of Bengal inspected the septic tank installation at the Fort Gloster Jute Mills, where a water-flushed latrine with 32 seats had been built on top of a septic tank, serving the needs of around 2,500 persons, and reported favourably about it. While admitting that the septic tank effluent did not reach the standards fixed for potable water, the sanitary officers reiterated the advantages of the new system over the 'intolerable nuisance' of the old one.[20]

Evidently, the mills resorted to the Hooghly to dispose of the large volume of effluents their septic tanks produced, an approach which by 1903 was starting to alarm the Indian, and more specifically, the Hindu population of the city. In early September 1903, the *Bengalee* published an article on the 'pollution of the river' through septic tank effluents from the mills:

> We understand that the managers of the river-side mills have hit upon an ingenious device for the disposal of night-soil. It is proposed to have septic tanks the contents of which are to be liquefied by some chemical [sic]

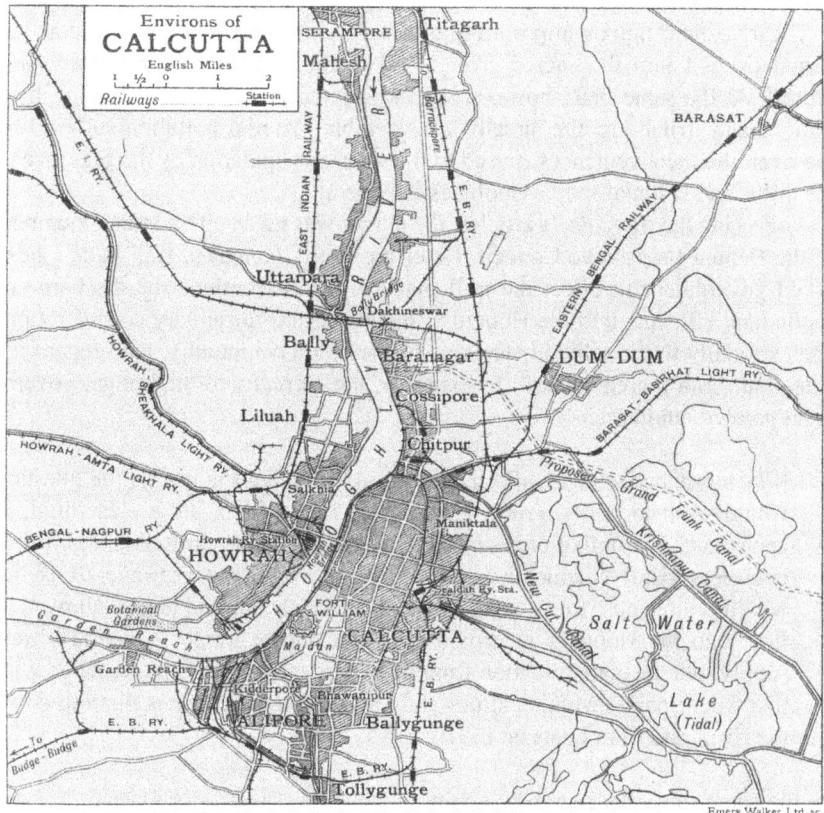

Figure 5.2 Calcutta and its Environs (source: J. Murray (1933) *A Handbook for Travellers in India, Burma and Ceylon, including all British India, the Portuguese and French Possessions and the Indian States*. 14th edn, London: Murray).

process and the liquid is to be discharged into the long-suffering river. How simple the method is to be sure! What a pity the idea should not have occurred to any of the mill managers earlier! Our rulers have done some very queer things in the name of sanitation but the proposed scientific pollution of the river which the Hindus regard as sacred, will, we are afraid, be an outrage upon orthodox sentiment. Thanks to the presence of the mills, the inhabitants of the riparian municipalities have already a great deal to put up with. But the proposed desecration of the river would, we very much fear, break the camel's back.[21]

Some days later, the *Bengalee* reiterated its protest, stating that while the discharge of septic tank effluents into the river was surely convenient for the mill-owners, it was 'an abomination to the Hindu Community with whom the

Bhagirathi is an object of devout worship and who bathe in their thousands every day in its waters'.[22]

Clearly, the religious impropriety of discharging what was conceived as liquefied excreta into the sacred river stood at the forefront of the *Bengalee*'s protest. At the same time, however, the article voices more long-standing frustrations and irritations: the already considerable material pollution suffered by the river, the inconveniences caused to the riparian population by the presence of the mills, and colonial sanitary policies in general.

Following the *Bengalee*'s articles, the matter was taken up by Indian members of the Bengal Legislative Council. Referring to the *Bengalee*, Kali Pada Ghosh asked government to direct the mill managers to discontinue the discharge of septic tank effluents into the Hooghly, 'having regard to sanitary considerations and especially to the religious feelings of the hindu community, who regard the Bhagirathi as a sacred stream'. In his reply, the secretary to the Bengal government pointed out that

> [t]he substance which is discharged from these tanks is not, as the question would seem to imply, crude sewage in a liquid form, but a clear fluid, in appearance not unlike ordinary river water. It is non-putrescible and nearly odourless [and] chemically quite different from crude sewage. It is not wholly innocuous, for it still contains bacteria, and could not be allowed to flow into the Hooghly in proximity to the intake for the Calcutta water-supply; but the contamination from this cause is insignificant compared with that from riparian Municipalities and villages, from carcasses floating down the river, and from boats on the river.[23]

At the same time, the secretary assured Ghosh of government's desire to keep the Hooghly 'as clean as possible' and informed him that the commissioners of the Calcutta and Burdwan divisions had been asked whether the septic tank effluents could be disposed of in another way.[24] While the Bengal government thus acknowledged the problematic bacterial contents in septic tank effluents, its answer clearly downplayed the controversial opinions on the issue among its own ranks. A couple of weeks earlier, the new sanitary commissioner of the provinces, F.C. Clarkson, had expressed his serious concerns over the discharge of septic tank effluents into the Hooghly. If the tanks were properly worked, the sanitary commissioner wrote, they would no doubt produce a clear, harmless effluent. But as his visit to a mill at Narayanganj had revealed, numerous tanks produced nothing short of crude liquid sewage, most probably due to overwork. The discharge of large amounts of such sewage into the river and its subsequent deposition on the river banks in his view posed a serious danger to the health of neighbouring populations. Consequently, he strongly advised that the 'indiscriminate use' of septic tanks be checked and that new installations be sanctioned only in special cases, until the reliable functioning of the new technology could be assured. In the case of existing installations, causes for failure and means for their prevention should be investigated. Additionally, the sanitary commissioner demanded the appointment of supervisors to keep track of the condition of septic tank effluents in relation to the number of people using the facilities, the amount of water used for flushing, and the number of hours given to each tank for rest.[25] Thus, in recommending further experimentation and strict supervision, the sanitary commissioners' assessment stood in strong contradiction to the earlier statement of the Bengal sanitary board, which had declared septic tanks to be a 'proved working success', having passed the experimental stage.[26]

In the meantime, Indian opposition against the mills' recourse to the Hooghly intensified and found expression through a wide range of newspapers. The *Sri Sri Vishnu Priya-o-Ananda Bazar Patrika* pointed out that the discharge of septic tank effluents into the Hooghly had 'greatly shocked Hindu feeling' and given 'mortal offence' to the community.[27] According to the *Sanjivani*, the 'religious feelings of the Hindus, who consider it a sacrilege to pollute the water of the sacred river Ganges with sewage', had been 'deeply wounded'.[28] And the *Bengalee* warned that 'the sanctity of the river which Hindus venerate as a goddess, is [...] threatened with destruction'.[29] The Ganges, it wrote,

> is with us Hindus an object of veneration. It is the receptacle of the ashes of the honoured dead. It is the emblem of purity. It is associated with thoughts of eternity, of endless birth in the endless ages to come, until the purified soul is absorbed in the Divine essence of Brahma. Such an object of veneration, linked in our minds with our holiest associations, we cannot with complacency allow to be polluted by loathsome discharges of resolved fœcal [sic] matter. We invite the authorities to view the question from the stand point of the great Hindu community.[30]

The wording used by the *Bengalee* drew on a public petition against the installation of a septic tank by the Rishra Hastings Mills for around 15,000 mill people, which was launched by Hindu inhabitants of Rishra and other localities in Serampore district with the support of the chairman of the Serampore municipality. In the eyes of the petitioners, the religious problematic was aggravated by the fact that the septic tank discharged its effluents right next to a large bathing *ghāṭ*, frequented by thousands of people to perform their daily ablutions and *pūjā*, and that this 'standing monument of public nuisance' itself was being built right in front of a popular Hindu temple. The petitioners, it was made clear, were not opposed to septic tanks per se, but wished to see them constructed at a more adequate site and their effluent discharged onto the adjoining fields or marshes to be used for irrigation and fertilising.[31] Indian resistance was not rooted in religious considerations alone however. The newspapers and the Rishra petitioners drew a direct connection between the discharge of septic tank effluents into the Hooghly and the prevalence of fever in Shamnagar, Nawapara and other localities where septic tanks had recently been set up.[32] To support their statement, the *Sanjivani* and others printed an extract of a report by the Calcutta health officer, Nield Cook. Asked to give his opinion on the construction of a septic tank at the new ordnance factory at Icchapur, Cook had warned that the discharge of the sewage of 2,000 to 3,000 people about one mile above the intake of the Calcutta water works would endanger public health:

> My reasons are that typhoid fever is by no means uncommon amongst the soldiers stationed in Barrackpore, that it has not been proved that typhoid germs are killed by the passage of sewage through a tank, and that I believe it possible for our water-supply to be infected. It has been proved that such germs can travel 50 miles or more in a river and retain their vitality; so the distance of one mile would not afford a sufficient protection.[33]

If the danger was such for the filtered water supply of Calcutta, how great then, the *Sanjivani* asked, must be the danger for people who drank directly from the river or bathed in it? Similarly, the Rishra petitioners called attention to the close proximity of the septic tank outfalls to the Howrah and Calcutta waterworks, and the fact that the townships of Rishra, Serampore, Mahesh and Chattra already suffered from outbreaks of cholera, smallpox and typhoid year after year. The case of the Rishra petitioners and others was prominently brought forward in the Bengal Legislative Council by Bhupendra Nath Bose, who demanded that government interdict the discharge of septic tank effluents into the Hooghly and direct their use for land irrigation.[34]

In reaction to the controversy, the Bengal government in February 1904 intervened. While upholding the superiority of septic tank technology over trenching, it acknowledged that the problematic of effluent disposal had not been given due attention, and that the present method was likely to have negative impacts on the city's water supply. Consequently, it passed orders to stop the discharge of septic tank effluents into the Hooghly, and directed the mill managers to consult the sanitary board as to the best means of effluent disposal.[35] The sanitary board, however, assessed the problematic very differently. 'It must be borne in mind', the board wrote in reply to the government's orders,

> that the filtrates are not sewage, but merely the water produced by liquefaction of solid organic matter during the process of putrefaction, together with the water used for flushing. These filtrates will not putrefy again. The Sanitary Engineer informs the Board that in appearance all the filtrates from the different mills are in all respects equal to those that he has seen in England.[36]

If the government feared secondary putrefaction, additional aerobic filters or filters filled with coarse material could be provided as a safeguard. But more than any filter, it was the self-purifying property of the Hooghly that could be relied upon. The large amounts of dissolved oxygen present in the river water were sure to rapidly oxidise and thus destroy the small quantities of organic matter that might find their way into the river. To underline its position, the board adduced an extract from a report issued by the Massachusetts State Board of Health, in which it was said that the danger from sewage discharge into a river was relative to the extent of its dilution, i.e. on the ratio between the river's water volume and the amount of sewage it received. Taking the formula applied in the extract, the sanitary board argued that in the case of the Hooghly, which it stated to have a dry weather discharge of 20,000 cubic feet per second, objectionable conditions could not be reached unless the sewage of more than three million people was put into it.

Simultaneously, the board sought to counter objections drawn from the highly unfavourable bacteriological composition of septic tank effluents. The purification of the effluents, it believed, could be easily effected by employing fine sand filters or cinder filters. But this, too, was of no pressing necessity, as the research conducted by the Massachusetts State Board of Health had proved that typhoid and *coli* bacteria got rapidly destroyed once they were exposed to sunlight.[37] Once again, the special character of the Indian environment – primarily the river's immense water volume and the 'tropical sun' acting on germs – were referred to as crucial factors intensifying the process of river self-purification.[38] The board's explanation is reminiscent of a passage in A.E. Silk's earlier report on the septic tank experiments at Presidency Jail. According to the sanitary engineer, the disposal of septic tank effluents in India was far less problematic than in Britain, due to the enhanced purificatory powers of large Indian rivers and the absorptive powers of their sandy beds in the dry season, and due to the sun and wind, which had much stronger purifying effects in India than in colder climates.[39] In conclusion, the sanitary board thought it to be expedient to allow 'the Hooghly to do its share of the purification work' and that letting it do so was preferable to calling 'upon the mill owners to incur additional expenditure in ultra-purification of the filtrates from their sewage works while the pollution of the Hooghly from other sources is so patent'.[40]

By early 1904, then, two major areas of conflict had emerged. The first one revolved around conflicting definitions of 'pollution'. Indian locals, newspapers and administrators who opposed septic tank effluent disposal into the Hooghly rested their resistance strongly – yet not exclusively – on a Hindu notion of 'pollution'. From this perspective, the discharge of what was considered to be liquefied excreta into the sacred Hooghly created a serious religious offense. British officials perceived this concept as incompatible with their own, wholly secular definition of 'pollution', which was guided by the discourse of contemporary Western science. At the same time, this secular concept of 'pollution' was in itself highly controversial, since different opinions existed about if and in what way septic tank effluents polluted river water, and, if it did, whether this pollution indeed posed a threat to the water supply of the capital. In view of the colonial government's further response to Indian protest, it is important here to emphasise the heterogeneity of Indian motives. As the various press reports and petitions that had emerged by early 1904 already suggest, local protest was not singularly based on religious motives, but was equally driven by serious concerns over public health, concerns that were based on the same scientific concepts of 'pollution' resorted to by the British.

A second area of conflict, somewhat less apparent at this stage, emerged along the lines of power. As the *Bengalee*'s very first article of September 1903 illustrates, relations between the mills and the Indian riparian population were already strained, and the septic tank controversy acted as a catalyst for the ventilation of these long-standing frustrations. Around March 1904, residents of the 24 Parganas district launched a petition against the installation of septic tanks in the Standard and Titagarh Jute Mills and several other mills at Khurda. In doing so, they pointed to the already considerable pollution of the river water caused by the mills' discharge of batching oil and soots into it, which was 'to the serious prejudice of the sanitation of the riparian villages'. If this existing pollution were compounded by that from septic tanks effluents, 'the situation [would] become an extremely serious one'.[41] As we shall see, such tensions acquired a definite political overtone during later stages of the controversy, when the Government of Bengal was accused of moulding its policy along the European mill managers' interests. Adding to the conflict's political potential was the fact that the protests involved some of the most prominent Bengali nationalists of the time. Thus, Bhupendra Nath Bose time and again raised objections in the Bengal Legislative Council, while Surendranath Banerjee's *Bengalee* counted among the newspapers reporting most regularly and vocally on the issue.[42] On the background of the tensed political climate of early twentieth century Bengal, it will be important to assess to what extent the septic tank controversy was woven into the overarching nationalist critique of colonial governance.

The Septic Tank Committee

In reply to the Bengal government's orders to advise mill managers on the disposal of septic tank effluents, the Bengal sanitary board expressed its inability to do so unless it was provided with more information. Three principal questions had to be answered, the board maintained, before it could execute its advisory role. First, government had to stipulate the distance above and below the Palta and Howrah water intakes within which the discharge of untreated tank effluents into the Hooghly was to be prohibited. Second, government had to decide whether it was sufficient to purify tank effluents by sand filtration before running them into the river. And third, government was asked to lay down a quality standard for effluents thus purified. Notably, these questions had been formulated by the chairman of the Indian Jute Mills Association during a special meeting of that body, which the secretary to the sanitary board, A.E. Silk, had attended to provide the mill managers with information.[43] In order to settle the questions raised by the sanitary board, the Bengal government in April 1904 decided to establish a special committee of inquiry, which in the following will be referred to as Septic Tank Committee. The Septic Tank Committee was directed to collect general information on the construction and situation of existing septic tank installations and the nature of the effluents they discharged. Moreover, it was to inquire into different methods of effluent disposal, namely land treatment, the use of effluents as boiler feed-water in mills, and their discharge into the Hooghly after sand filtration. In the latter case, the Committee was to state the conditions under which effluents could be discharged into the river without any detrimental effects on public health. Last but not least, the Committee was to answer the specific questions put in front of government by the sanitary board. The membership of the Committee, significantly, was all-British: it was presided over by S.H. Browne, inspector-general of civil hospitals, Bengal, and had two members, Bengal's sanitary commissioner F.C. Clarkson and its chief engineer D.B. Horn. L.P. Shirres, secretary to the municipal department of the provinces, served as secretary.[44]

In December 1904, the Septic Tank Committee issued its report.[45] It was labelled as an interim report, since experiments on the bacteriological purification of effluents through sand filtration and the definite assessment of the Hooghly's water volume during the dry season were still in process. Nevertheless, certain definite conclusions had been arrived at, which the Committee thought to be ripe for communication due to the primary importance of the matter. Before proceeding to a more detailed analysis of certain aspects of the report, its main contents will be summarised here to serve as a foundation for the further discussion: Part I of the report explains the backgrounds of the Committee's appointment and the objects and methods of its inquiry. Information had been collected through questionnaires circulated among mill managers, through personal inspection of septic tank installations and effluents, and by collecting oral evidence from various Hindu 'witnesses'. Moreover, chemical and bacteriological examinations had been taken of tank effluents, and of river water near points of discharge and near water supply intakes. As a further source of information, the reports of the British Royal Committee on Sewage Disposal and other relevant literature had been consulted.[46]

Part II contains the Committee's conclusions on the discharge of untreated septic tank effluents into the Hooghly, as hitherto practised by the mills. Visual and olfactory assessments of the effluents had overall proven satisfactory, and the non-putrescibility of the effluents was assured. The chemical and bacteriological examinations, however, had furnished very unfavourable results. As indicated by the low amounts of nitrates and the large amounts of albuminoid ammonia present in the effluents, the chemical oxidation of organic matter was incomplete. The results gained by bacteriological analyses were even less promising. Some pathogenic bacteria like the *B. enteritidis sporogenes* were entirely or almost absent from the effluents, but they held a large number of *coli* bacteria comparable to that found in ordinary crude sewage, and also contained a certain number of cholera bacilli. Consequently, the river water around the outfalls was found to be considerably polluted, even though this pollution had largely disappeared around the Palta water intake.[47] The problematic bacterial content of the effluents led the Committee to the conclusion that they indeed posed a potential danger to the capital's water supply – not during most of the year, when the dilution and self-purification effected by the large water volume of the Hooghly 'in all probability' provided a sufficient safeguard, but during the dry months, when the Hooghly's fresh water discharge was estimated to be as low as 2,000 cubic feet per second (and thus similar to British rivers). Therefore, the Committee advised that the discharge of untreated septic tank effluents be entirely stopped above the water intakes and allowed only after Tolly's Nala.[48]

Part III of the report discusses different methods of effluent treatment. As to the application of effluents for land irrigation, the Committee made out several disadvantages. Generally, the soils in the neighbourhood of the Hooghly were fertile and in no need for much fertilisation. Moreover, during the wet season they would not be capable to absorb large quantities of effluent, which would lead to oversaturation of the land and the leakage of the effluents into surrounding tanks used for drinking water supply. Thus, the character of the soil in the vicinity of the Hooghly and the general climatic conditions precluded the use of effluents for land irrigation. Much more optimism was expressed about their use as feed-water for boilers in the mills. This was practiced in certain mills in Britain, and had proven to be without problems. However, doubts were expressed by mill managers, who feared that the effluent could affect the boilers, and, more importantly, that their staff may be reluctant to come into contact with the effluents for religious reasons. The men cleaning the boilers were mostly Hindus, who, the managers apprehended, quite likely object to enter boilers in which septic tank effluent had circulated. According to the Committee, this could be easily overcome by hiring a *mehter* for the task. Lastly, the bacteriological purification of septic tank effluents by sand filtration was considered to be a most suitable method by the Committee. As per the extent of purification to be reached, the Committee suggested that the degree of effluent purity should comply with the water quality of the Hooghly at the very point of effluent discharge. But as experiments were still in the process, a detailed discussion and evaluation of the issue was considered premature.[49]

In the fourth and last part of the report, the Committee wrote its recommendations for improving technical aspects of the tanks, such as the flushing system or problems derived from overworking. Overall, the Committee acknowledged that problems existed, but strongly commended the continued use of the tanks by the mills. Septic tanks, the Committee noted, marked a 'great advance in Sanitary Science', and even though they presently caused some pollution, the latter was likely to increase if the septic tanks were stopped.[50]

The Septic Tank Committee report as such is rather short and its conclusions technical. However, its extensive appendix affords important insights into the Committee's general approach towards Indian opposition, and into the ways in which it confronted and negotiated Hindu religious notions of 'pollution' in particular. The following analysis demonstrates that the Committee primarily focussed on the religious component of Indian opposition, and that it pursued a clear-cut agenda to establish a definition of ritual 'pollution' that would allow it to negate the validity of religious opposition altogether.

The Committee called upon several Hindu members of the city's elite to express their views on the discharge of septic tank effluents into the Hooghly, and to explain their and the general public's motives for dissent. The witnesses consulted can be divided into two groups. The first group consisted of men living in localities where septic tank effluents were being discharged into the Hooghly or one of its tributaries, and/or who had raised their voices against river pollution prior to or during the septic tank controversy: Raja Peary Mohun Mookerjee and Rai Abinash Chandra Banerjee from Bally, Baman Das Banerjee from Rishra, Nalin Bihari Sircar and Narendra Nath Sen.[51] The second group, to which we will turn at a later stage, consisted of religious 'experts', whom the Committee consulted to obtain an 'authoritative' opinion on the matter: Yogisa Chandra Sastree, professor of Sanskrit at Doveton College and examiner at the University of Calcutta, Pundit Kaliprosanna Bhattacharya and Pundit Satis Chandra Vidya-bhusana, both professors of Sanskrit at Calcutta's Presidency College, and Mahamahopadhyaya Raj Krishna Tarkapanchanan and Pundit Rajani Kanto Vid-yaratna, both pundits from Nadiya.

In the course of the investigation, the Septic Tank Committee and the Indian witnesses – both the 'lay' witnesses and the religious 'experts' – came to divide the motives for their opposition into three distinct categories: sanitary, religious and sentimental. This categorisation, as we shall see, proved crucial for the Com-mittee's response to Indian/Hindu opposition and the related conclusions com-municated in its report.

As to the sanitary side of the question, all 'lay' witnesses unanimously thought of the effluents as a serious menace to public health. As Narendra Nath Sen pointed out, the effluents not only threatened to compromise the quality of the city's water supply, they also directly affected the many people – both Hindus and non-Hindus – who regularly took bath and drank from the river near the mills. This was especially true for the large numbers of coolies who lived in their neighbourhood. The fact that the number of mills was increasing year by year enhanced the urgency of the matter.[52] As a solution, Sen and others strongly argued for the disposal of the effluents on land, or, in cases where this was found impracticable, their use as feed-water for boilers.[53] What stands out is that com-plaints about the sanitary problematic of septic tank effluents were intricately linked to complaints about the discharge of manufacturing wastes by the mills, which in the witnesses' eyes contributed greatly to the pollution of the Hooghly and its smaller tributaries. Fearing the further deterioration of the Bally creek through septic tank effluents, Abinash Chandra Banerjee noted:

> When the [factory] effluent is thrown into the khal, we cannot bathe in the river [...]; such an amount of noxous [sic] smell is emitted from the water. Some times [sic] we have to draw the water for drinking purposes, and we cannot use it even after purifying it by some process.[54]

In his report on existing septic tank installations in the Serampore district, the chairman of that municipality, Kisorilal Goswami, wrote:

> It is generally believed, and I share that belief too, that the discharges from the riparian mills, especially those from the paper and bone mills into the River Hooghly between Calcutta and Gariffa on its one side and Howrah and Chandernagore on the other have contributed in a substantial measure to

the unhealthiness of the people the great majority of whom use the river water for all purposes. Not being a medical expert my opinion is not founded on any scientific analysis of the discharges from the paper and bone mills and the septic tanks, but so far as their colour and smell would indicate they seem to me to be very repugnant indeed.[55]

The effluents of the paper mills most likely included carbonate of soda and lime as well as the wash of bleaching powders and old rags, the usual waste products of this type of mill.[56] Bone mills produced bone meal, which was used as an agricultural fertiliser. To obtain the meal, animal bones had to be boiled in order to make them brittle and to skim off the fat. They then were reduced to small pieces either by chopping or by being put through toothed cylinders, and were finally ground into powder by millstones.[57] It is easy to believe that the bone mills referred to by Kisorilal Goswami produced rather unpleasant waste products and smells, the latter being referred to by the *Howrah Hitaishi* as a 'powerful stench'.[58]

The most elaborate criticism about factory wastes came from Nalin Bihari Sircar. As a member of the Calcutta Corporation, Sircar in the early 1890s had lodged a complaint against discharges from the Titagarh and Kankinara paper mills, which he believed compromised the safety of the Palta and Howrah water intakes.[59] This episode is important for our further discussion, as it reveals how the Bengal government handled river pollution by factory wastes *before* this was prominently discussed in the context of the septic tank controversy, and, more specifically, to what extent mill managers were able to direct related government policies.

Acting on the Corporation's complaint, its health officer W.J. Simpson in 1894 investigated the alleged pollution of the Hooghly by the Titagarh and Kankinara paper mills. According to Simpson, the effluents were of a very impure and objectionable character and endangered the city's water supply. His case was however weakened by the fact that regular chemical analyses of river water had been initiated only in 1892, one year before the Kankinara mills started operating. It was therefore impossible to compare the present water samples with samples of previous years over a longer period of time, which left him with inconclusive results. Despite this scanty evidence, Simpson presented his case as one of principle, and one with great implications for the future:

> The question [...] is not altogether one whether there is proof of disease or liability to disease produced in Calcutta or among the inhabitants who take their drinking water from the river, and who live in much nearer proximity to the mills, but there is also the question whether the mills shall be allowed to use the river as a sewer, and discharge any objectionable matter which may be convenient into it. I take it that the river is used because it is a simpler and cheaper mode of disposal of the waste water of the mills than the construction of proper drains and the purification of this water on lands in their own districts, and if one or two mills are to be allowed this

privilege, it is difficult to know on what grounds other mills may be dis-allowed, and there can be no doubt that such a multiplication would be at the expense of the public health of all the inhabitants who may have to take their drinking water from what must ultimately become, notwithstanding the volume of water, a most potent cause of ill-health.[60]

Simpson's argument evidently convinced the Bengal government. Shortly after receiving his report, Lieutenant-Governor Charles Elliot informed the Bengal Chamber of Commerce of his intention to introduce special legislation with the aim to regulate the discharge of mill and other factory effluents into the Hooghly. The putative bill, Elliot explained, would be applicable to all mills and factories situated along the banks of the Hooghly, and

> would contain a prohibition against persons who should cause or allow to be discharged into the river any solid or liquid sewage matter, or any poison-ous, noxious or polluting matter, whether liquid or solid, proceeding from any factory or manufacturing process.

Existing mills would be given six months' time to arrange for the alternative dis-posal of their waste, or to render it innocuous. Where the defendant was a company or a manager of a mill or factory, prosecution may only be taken up with the sanction of the commissioner of the respective division or the Lieutenant-Governor himself. This projected concession to industrial interests in specific, and the contents of the draft in general, are strongly reminiscent of the clauses of the British Rivers Pollution Act of 1876, and it seems more than likely that Lieutenant-Governor Elliot in drafting the bill had the latter's provisions in mind. Anticipating resistance against government interference, Elliot foreclosed the range of arguments most likely to be brought forward by opponents against legislation: That the amount of waste discharged was only little, that few mills and factories used dangerous chemicals, and that the water volume of the Hooghly was great. But following Simpson, he pointed out that

> the facts alleged make it easier to intervene now when but few vested inter-ests will be affected: after some years, when the number of mills has increased and the mischief is greater, the difficulty of interference will be largely enhanced.

The Lieutenant-Governor's letter was sent to the Chamber of Commerce, invit-ing criticism and alternative proposals on how to secure the protection of Calcut-ta's water supply. At the same time, government called upon the special inspector of factories to submit a report on the number and nature of mills and factories situated at or near the river banks, their methods of waste disposal, and possible complaints received about the latter.[61]

The reply issued by the Chamber of Commerce was short and crisp, and exactly followed the course anticipated by the Lieutenant-Governor. The

Chamber found that 'looking to the comparatively small quantity of objectionable matter passing into the river, and to the immense volume of water passing down its bed' it could not 'but come to the conclusion that there is no necessity for any special legislation, and that a resort to such legislation would be a matter for extreme regret'. The danger posed to the city's water supply being negligible, the Chamber saw no need for proposing any alternative measures for its protection.[62] The Chamber received significant backing from the inspector of factories. According to his observations, the jute and cotton mills as well as the workshops and dockyards discharged little or no waste into the river, neither solid nor liquid. The Titagarh, Imperial, and Bally paper mills, he noted, discharged around 700,000 gallons each of a 'brownish-coloured' liquid into the river, containing predominantly the washings of rags and grass, but also smaller amounts of lye, alum, lime and chlorine. While the lye (used in the washing of the rags) was recovered by evaporation, the alum, lime and chlorine discharged into the river water, according to the inspector, had a beneficial effect on the latter, as he believed them to be lethal for germs and to precipitate solid matter. Moreover, he was assured by the manager of the Imperial Mill that the quantity of effluents the mill discharged did not exceed that of a paper mill he had managed in Scotland during earlier years. Most importantly however, the inspector observed that the suction and intake pipes of all the three mills lay roughly 100 yards from each other, from which he followed that the refuse water was factually reused by the mills themselves. Thus, the inspector concluded that the introduction of legislation was unnecessary. However, he suggested that the managers of the paper mills be requested to evaporate the washings in order to reduce the amount of liquid discharge, and to make use of incinerators as much as possible.[63]

Based on the reply received by the Bengal Chamber of Commerce and the report of the inspector of factories, the Bengal government decided to drop its plans for introducing legislation. Looking at the initial enthusiasm of Lieutenant-Governor Elliot and the evidence on which the decision was taken, this move appears rather peculiar. In the process, the Lieutenant-Governor's main point – the timely control of a source of pollution that in future was bound to increase significantly – was literally ignored by both the Chamber and the inspector of factories, and the evidence on which the decision rested was based either on visual observation or furnished by the mill managers themselves. No steps were taken to ascertain the actual amount of effluents discharged, their degree of harmfulness, or their effect on the quality of the water taken in at Palta. Even though no direct evidence is at hand, it may well be justified to see this as a successful intervention of Calcutta's European industrial lobby. As Dipesh Chakrabarty has observed, the Bengal government was closely united with owners of capital, especially those represented by such powerful organisations as the Bengal Chamber of Commerce and the Indian Jute Mills Association. This 'natural unity' found its fullest expression during the nineteenth century and carried well into the twentieth century. Mill owners and managers shared close social relations with colonial officials, and they often served as bureaucrats themselves, for example in municipal committees or as honorary magistrates, and as such were an influential element of Calcutta's colonial ruling-class. Accordingly, the Bengal government closely shielded the mill managers' interests during labour unrests and when it came to the introduction and supervision of factory laws. Moreover, inspectors of factories tended to report altogether favourably on the conditions found in mills, such as the work and living conditions of the mill-hands or the arrangements for medical care provided to them. In some cases, the Bengal government even actively discouraged factory inspectors from including 'controversial' information in their reports, or they were altered after successful interventions by mill managers.[64] Seen in this light, the Bengal sanitary board's repeated favourable statements about the mills' septic tank latrines can be interpreted as an expression of its close affinity towards industrial interests.[65]

Thus, in 1895, Calcutta's mill managers had successfully influenced the Bengal government's river pollution policy, preventing it from introducing legal measures aimed at pollution prevention and control.

Informed by the Septic Tank Committee that government in 1895 after 'careful analysis' (which, considering the above, is obviously an overstatement) had found no pollution at all through manufacturing wastes, Nalin Bihari Sircar retorted:

> That may be so. But this much I know as a fact, that ever since people have been bitterly complaining of discharge of refuse matter from the mills. They complain of oily black scum, emitting very offensive smell, discharged from the mills, that floats all over near the river bank and sticks to the body of those who come to bathe in the river. This is no doubt very disgusting and objectionable [...] both on sanitary and sentimental grounds.[66]

The problem existed elsewhere, too. The manufacturing wastes of the Gouripore Mills at Naihati, according to Sircar, all flowed into a narrow creek that had formed between the Hooghly and its permanent bank, due to the increasing siltation of the river in that area. Being blind at its north end, the water of the creek during flood tide was pushed upwards and left the whole waste matter of the mills floating on its surface, much to the annoyance of the people who were forced to cross the creek in order to reach the Hooghly. What made things worse was that the mill's effluents carried a good amount of excreta and urine from the mill's latrines and urinals with them. Here too, Sircar strongly suggested that the Gouripore Mills remedy the situation by disposing their effluent on nearby land.[67]

While sanitary considerations thus formed an essential part of Indian opposition – Narendra Nath Sen even made it clear that his objections were *entirely* based on sanitary concerns, and were not motivated by any religious or sentimental reasons –,[68] the Septic Tank Committee was not so much interested in its witnesses' views on sanitation and public health than in their religious perspectives. Except for Narendra Nath Sen, all witnesses explained that their opposition also had strong religious motives. Asked to define these, Raja Peary Mohun Mookerjee replied: 'Our religion says "Apa Narayan", i.e. water is God. We say rivers are goddesses. Not even the most ignorant people would spit into a tank or any water. It is God itself.'[69] Similarly, Nalin Bihari Sircar stated:

> [T]here are very clear and distinct injunctions in the Shastras against pollution of water used for domestic purposes. In Manu Sanhita it is laid down that water should not be polluted by making water in it or defecating therein, or throwing phlegm into it, or by washing therein cloths smeared with fecal matter, etc.[70]

Baman Das Banerjee's reply in part reproduced a paragraph contained in the Rishra petition:

> [T]he river is held most sacred by the Hindus. It is an object of veneration and worship. It is the receptacle of the ashes of the honoured dead. It is the emblem of purity and is associated with thoughts of eternity, of endless birth in the endless ages to come, until the purified soul is absorbed in the Divine Essence of Brahma. Such an object of veneration, linked in our minds with our holiest associations, should not be polluted by loathsome discharges of resolved fæcal matter. The pollution of the river is viewed by all Hindus as a great sin. As to the isolated instance of committing nuisance stealthily on the river by a few low class men and boatmen, the practice is not at all allowed in the Shastras, which call it a great sin.[71]

At a closer look, these statements are statements about *acts* of pollution, namely, acts of putting a polluting substance into the sacred Ganges. This, it was pointed out, had been strictly prohibited by different Hindu scriptures, particularly by the Manu Samhita. However, in what was a meaningful twist, the Septic Tank Committee tried to lead its witnesses into expressing their opinion on what *impact* such acts had on the river's inherent sacredness. The following exchange between the Committee and Baman Das Banerjee is exemplary for the Committee's strategy:

QUESTION [Q]: You say that you object to the sewage going into the Ganges because it pollutes the water. Is that pollution in the physical sense or in the religious sense?

ANSWER [A]: Of course I am not a medical authority or expert. So my objections are more in a religious sense than in a sanitary sense.

Q: Is not the Ganges a religious river?

A: Yes.

Q: Can the Ganges be polluted in a religious sense?

A: Practically it can. As I am not well versed in the Shastras I cannot say that.

Q: What do you think your Pandits will tell us if we consult them? And if they tell us that the Ganges being a holy stream it cannot be polluted, would you object to that?

A: There are sanitary, sentimental, and religious objections.

Q: But if your Pandits tell us that the running water of the Ganges cannot be defiled by impure matters being discharged with it, won't you accept that opinion?

A: Of course we might accept it, but not to our heart's content, to be frank.[72]

The Committee's exchange with Abinash Chandra Banerjee was led in a similar vein:

Q: Do you mean to say that the river is to be defiled from a religious point of view?

A: Yes, it is. Even in the Manu Sanhita we find that no one should commit any nuisance in the river.

Q: It is not an ordinary river. Don't you think the Ganges can never be polluted?

A: That is impracticable […].

Q: What do you say about the droppings from the boats, etc.?

A: An orthodox Hindu will never commit any nuisance from the boats.

Q: But we have seen some.

A: That is sometimes done by boatmen, etc. (some ignorant people), and that we cannot help. It cannot be stopped by anything but legislature.

Q: Is that from a religious view?

A: Yes it is. […]

Q: Would you be surprised to hear that some of your pundits hold that the running water of the Ganges cannot be polluted or defiled?

A: I am not competent to say that, as I am not well versed in Sanskrit and Shastras […].[73]

Thus, all the witnesses considered the *act* of putting a polluting substance into the Ganges as a serious religious offense, prohibited by Hindu scriptures. The majority further believed that the sacred purity of the Ganges indeed got affected by the discharge of septic tank effluents into it. This was most clearly maintained by Baman Das Banerjee, while Abinash Chandra Banerjee's statements appear somewhat inconclusive. Banerjee's view was shared by Raja Peary Mohun Mookerjee, who believed that the sacred water of the Ganges could not be polluted 'in the long run', but that 'immediate pollution [was] possible in the religious sense'.[74]

With these two principal religious arguments at hand, the Septic Tank Committee undertook an obvious attempt to define ritual pollution in a way that would invalidate the religious opposition brought forward by Mookerjee and others. Its persistent allusion to 'some of your pundits',[75] who apparently held that the Ganges's purity remained unaffected when coming into contact with septic tank effluents, points towards the Committee's strategy of how to do this. The pundit referred to was Yogisa Chandra Sastree, professor of Sanskrit at Doveton College and examiner at the University of Calcutta, whom the

Committee seems to have consulted before taking the opinions of its 'lay' witnesses. While interviewing Sastree, the Committee without any detours turned straight to its primary point of interest:

> Q: Do you think that according to Hindu Sastras, any sewage, if thrown or flows [sic] into the Ganges, will defile its water in a religious sense?
>
> A: I do not think that according to Hindu Sastras the water of the Ganges will be defiled in a religious sense, if any sewage is threwn [sic] or flows into it. The Ganges water is pure in itself, and therefore it cannot be defiled or be impure by any means whatever, while it is in current. But when it is kept in any pot it may be defiled in certain cases, such as, if it is kept in a pot in which wine was kept before […].
>
> Q: What are the points of objection to the sewage flowing into the Ganges?
>
> A: It is very disgusting to bathe in such water or drink it. I was told that the effluents are injurious to health […].
>
> Q: Then it is clear that the Ganges water cannot be polluted in a religious sense?
>
> A: Yes, it is clear that the Ganges water cannot be polluted in a religious sense while it is in current.[76]

To fortify his position, Sastree drew up an elaborate note. After giving textual evidence of the sacred character of the Ganges and its spiritual powers, the pundit listed the 13 prohibited acts in relation to the sacred river laid down in the Brahmanda Purana.[77] However, Sastree pointed out:

> It should be here borne in mind that the above prohibitions of the *Brahman-dapuranam* do not mean to say that the water of the Ganges shall be defiled by those acts. The wise sages have enjoined some of them simply to keep the Ganges water free from dirts and some for inducing firm devotion to the Ganges; but not at all to protect it from defilement. According to Hindu notion, dirty water and defiled water are two widely different things, inasmuch as the latter can never be used for the purpose of worshipping the deities or offering oblations to ancestors. […] For these reasons we must understand from these prohibitions that if anybody disobey them, he shall be a sinner of commiting [sic] forbidden acts, and shall be liable to expiation; but his acts can never pollute, defile or make impure or unholy the water of the Ganges.[78]

Sastree acknowledged that the Dharmashastras actually contained no *direct* information on whether or not excrements caused the pollution of the Ganges itself, in the sense of ritual defilement, and that he had reached his conclusions by a series of deductions, drawn from two passages in the Manu Samhita and one in the Vishnu Purana. The Manu Samhita, as we have seen, contains a prohibition against the throwing of excremental matter into water; some sections later, it states that 'a river is purified by its current'.[79] Similarly, the Vishnu Purana, according to Sastree, stated that 'a river is purified by its current carrying away all slime and mud'.[80] Taking these passages, Sastree theorised that ordinary river water could indeed be polluted by excrements in a physical as well as a ritual sense, but that the water quickly reverted to its original state of purity thanks to the purifying action of the river's current. Accordingly, ordinary river water could never assume a condition in which it became unfit to be used for religious rites, as it was immediately purified by its own currents. What was true for ordinary river water was even more so for Ganges water. To Sastree, a river so sacred could never be subject to defilement, and as no scripture listed any rituals for the purification of Ganges water it could be followed that such a defilement according to the scriptures did not exist. Even if a possibility of the defilement of Ganges water were admitted, the rule 'a river is purified by its current' applied and guaranteed the rapid purification of the sacred river.[81]

On a more practical level, Sastree pointed to the fact that Hindus kept doing their rituals in the Ganges even while knowing about the physical pollution committed by people defecating along rivers banks and from boats, which proved that Hindus did not believe in a possible defilement of Ganges. Summing up, Sastree wrote:

> Under these circumstances, I am of opinion that the Ganges water, while in current, cannot be defiled by any means whatever, and no objection can be raised from the religious point of view if any sewage is thrown or any sewage elffluent [sic] flows into it.[82]

With Sastree's evidence in hand, the Septic Tank Committee consulted four more scholars and pundits: Pundit Kaliprosanna Bhattacharya and Pundit Satis Chandra Vidyabhusana, both professors of Sanskrit at Calcutta's Presidency College, and Mahamahopadhyaya Raj Krishna Tarkapanchanan and Pundit Rajani Kanto Vidyaratna, both pundits from Nadiya. Their testimonies reveal perspectives rather different from Sastree's. Other than him, all but one believed that Ganges water could at least temporarily be polluted in a religious sense, or polluted so as to make its ritual use impossible.[83] Significantly, the particular character of septic tank effluent was believed to have special implications in this regard. Whereas with solid pollutants – such as dry excreta – the area of pollution formed a clearly definable space, the liquid effluent merged with the river water and precluded any process of self-purification through the current.[84] Furthermore, all of them strongly stressed the prohibition against any act of throwing ritually impure matter into the Ganges, and into water bodies in general, referring to the injunctions found in various Hindu scriptures. Quoting from the Brahmanda Purana, the Vedas and other texts, Kaliprosanna Bhattacharya maintained that

> any act of irreverence done towards the sacred river is considered by the Hindus as an act of *sacrilege*, and, therefore, any organised system of throwing in objectionable matters into it [...] is most repugnant to all sections of the Hindu communities on religious grounds.[85]

But again, the Septic Tank Committee was primarily interested in obtaining statements about whether or not the Ganges itself could be defiled by septic tank effluents, and conducted its interrogations accordingly. The following interview with Raj Krishna Tarkapanchanan aptly illustrates this:

Q: What is your name?

A: Raj Krishna Tarkapanchanan.

Q: Where is your residence?

A: Nawadwip.

Q: Can Ganges water be polluted from a religious point of view?

A: In our shastras, the discharge of any filthy substance into the river is strictly prohibited, as we have to use the water for bathing, cooking and drinking purposes.

Q: Can Ganges water be polluted by the discharge of any filth from a religious point of view?

A: Ganges being pure in itself, it cannot be polluted in a religious sense, but so long as we can see or know that some sort of nuisance is committed into it, or any filth is floating on the water, or that the sewage effluent is discharged into the river, we cannot do any religious duties with that water; we cannot offer that water to our Gods with our full heart.

Q: Can Ganges water be polluted in a religious sense?

A: No, it cannot be.[86]

In sum, Sastree's viewpoints were in no way unanimously shared by his colleagues. All of them agreed that the discharge of excremental matter into the Ganges represented an act of gross irreverence, strictly forbidden by some of the most authoritative Hindu scriptures. More importantly, several of them disagreed with Sastree on the question whether the Ganges water itself got defiled in some way or other by coming into contact with septic tank effluent and rendered unfit for ritual purposes. Thus, while the Septic Tank Committee presented Sastree's conclusions as authoritative and representative for pundit opinion, it was actually highly contested.

The third category used to classify the motives for Indian opposition, used by both the Septic Tank Committee and the witnesses themselves, was the category of 'sentiment'. While the religious category was defined by recourse to Hindu scriptures, 'sentiment' referred to emotional and mental sensations. Both the 'lay' witnesses and the religious 'experts' agreed that from a sentimental point of view, the discharge of septic tank effluents into the Ganges was highly objectionable. According to Raja Peary Mohun Mookerjee, the practice was simply 'abhorrent to our feeling'.[87] Yogisa Chandra Sastree, too, thought that it was very disgusting to bathe in or drink water into which sewage had flowed.[88] Moreover, Satis Chandra Vidyabhusana explained: '[i]t will certainly be repugnant to our feelings on the sentimental ground to imagine that the water to which we are praying and the water which we are presenting to our forefathers are mixed with the discharge of latrines'.[89] Raj Krishna Tarkapanchanan even more clearly perceived the sentimental question as inextricably linked to the religious one in the context of religious practice. The pundit was convinced that Ganges water in itself could not be defiled through the contact with sewage. However, he believed that it could become impure in the sense that its offensive condition prevented devotees from performing their religious worship wholeheartedly, thus affecting their religious practice and lessening its benefits. Therefore, Tarka-panchanan believed that Ganges water was impure as 'long as the filth is within our sight and knowledge', 'but when it is out of sight or beyond our knowledge we do not think the water to be impure'.[90]

The significance of the distinction between 'religious' and 'sentimental' objections and the Septic Tank Committee's strategy to determine the validity of specific religious definitions of 'pollution' becomes apparent in the Committee's report, where it summarised its conclusions gained from the interrogation of Hindu witnesses in a short paragraph:

> The sentimental and religious objections to the pollution of the Hooghly have not been discussed, but it will be seen from the minutes of evidence that a difference of opinion exists regarding the latter, and the real objection appears to be sentimental rather than religious, and to be largely due to ignorance of the transformation effected in the nature of the sewage by its passages through the septic tanks. [The Committee] conclude[s] therefore that the opposition would probably subside if the real facts were brought home to the public, so that the great utility of the new system might be recognised.[91]

Summing up, the way in which the Committee negotiated the religious concept of 'pollution' leaves little doubt that its primary aim in investigating Hindu witnesses was to invalidate religious opposition. The Committee deliberately over-looked its witnesses' repeated reference to the injunctions of the Manu Samhita and other scriptures against *acts* of pollution, and narrowed the religious argument down to a scripture-based discussion on whether or not the Ganges itself could be defiled by septic tank effluents. It then selectively treated Sastree's viewpoint as authoritative, even though it was contested by other pundits, and used it to deny the validity of religious opposition altogether.

Second, by concluding that 'the real objection appears to be sentimental rather than religious', the Committee also refused to attach any value to actual religious practice and feeling, equating religious feeling with mere 'sentiment' on the one hand, and ignoring arguments about the interconnectedness of religious practice and sentiment, on the other. Instead, the Committee upheld the scriptures, together with their 'expert interpreters' the pundits and scholars, as the sole source of religious authority. In this it followed a long-held colonial practice, established by British Orientalists working in India during the late eighteenth and early nineteenth centuries, which ascribed absolute authority and authenticity to allegedly classical Hindu texts, while denigrating contemporary religious practices as impure or debased. Ancient India, the idea went, had been the 'Golden Age' of Hindu culture, and contemporary Hindu practices were mere corruptions of the latter. This Orientalist trope – which was readily used to justify the colonisers' interest in India, as well as their project of restoring its 'civilisation' – was in turn adopted by some of the most influential Indian intellectuals and socio-religious reformers of the nineteenth and early twentieth centuries, such as Rammohan Roy and Swami Vivekananda, and became one of the most pervasive and significant ones in shaping modern Hindu thought.[92] In this light, the 'lay' witnesses' lack of confidence to express themselves authoritatively on the religious question, due to their insufficient knowledge of the scriptures, is noteworthy.

The Committee's orientalist bias is equally borne out by its claim that Indian opposition stemmed from a lack of understanding of how sewage got transformed through the septic tank process. This depiction of the Indian public as uninformed about contemporary science stands in obvious contradiction to the arguments brought forward by Indian newspapers and the Committee's witnesses. By the early twentieth century, Bengal's Western-educated elite had long recognised the legitimacy of modern Western science,[93] and the press reports and interviews clearly show that they were very much aware of the scientific aspects of water quality and the latter's relation to disease. An article written in the *Bengalee* in August 1904 moreover suggests a keen awareness of current events in Britain. The Royal Commission on Sewage Disposal, the *Bengalee* pointed out, so far had not reached any definite conclusions as to the safety of sewage disposal into rivers and the methods that could neutralise harmful bacterial contents. Therefore, the *Bengalee* argued, it was wrong to simply assume that the discharge of septic tank effluents into the Hooghly was harmless, and the river's comparatively great volume was 'no argument in favour of a relaxation of the rigid rules of hygiene in the case of India'.[94] That said, the *exclusively* scientific perspective did indeed not resonate with the Hindu majority, which viewed septic tank effluents as nothing other than 'noxious human excreta' 'in spite of its transformation', that had lost nothing of their ritually polluting character.[95]

Unsurprisingly, the Septic Tank Committee report drew severe criticism from the Indian press. The Committee's work had been followed closely over the months, and even before the publication of its interim report in late December 1904, newspapers had repeatedly questioned the trustworthiness of its proceedings. The strongest criticism was levelled against the Committee's apparent complacency about a number of mills which, against government orders, continued to discharge septic tank effluents into the Hooghly while investigations were still in process.[96] According to the *Bengalee*, the Committee even gave special permission to the Hastings mills at Rishra to put their septic tank to work again.[97] In October, Baman Das Banerjee and other Rishra residents forwarded a petition to the Septic Tank Committee president, requesting that the discharge of effluents be stopped for the time of Pitra Paksh.[98] However, a couple of days later, Banerjee complained that things had turned from bad to worse, as the mill manager – allegedly in vengeance for the sustained agitation – had started to discharge the effluents during the early morning hours, exactly during the time when the majority of people performed their rituals in the Hooghly. According to Banerjee, the river bore such an abominable stench that Hindu widows had to retreat from it, unable to perform their ritual bath. Commenting on this account, the *Bengalee* insisted on the prompt intervention of government, and, on a sharper note, concluded:

> Surely this is not affording that religious protection to which the people are entitled and which is the declared policy of the Government. Are the religious interests of the people to be sacrificed for the sake of British capital? We pause for a reply.[99]

Allusions towards the Bengal government's alleged partiality, based on race and shared commercial interests with the mill managers, were repeatedly made. To the *Sri Sri Vishnu Priya-o-Ananda Bazar Patrika*, for instance, it was clear that government tolerated river pollution through factory wastes because the factory owners were European. And the *Bengalee*, in its very first article of September 1903, expressed its doubts about the success of the Rishra petition, as 'experience has unfortunately convinced [people] that Government, like Providence, always takes the side of the heavy battalions which are in the present case represented by the mill managers'.[100] Taking into account the close ties between British capitalists in Calcutta and the Bengal government, this reproach was certainly not without foundation. Some evidence exists which clearly indicates that these ties directly influenced septic tank policy. When residents of Telinipara petitioned the Bhadreshwar municipality against the continued discharge of septic tank effluents by the Telinipara mills, the British chairman of the municipality refused to comply with the request, stating that he was not aware of any government orders. Significantly, he was occupying a leading position within that very mill.[101] After cholera broke out in and around Rishra in November, the press denounced the Bengal government for its failure to enforce the temporary stoppage of septic tanks. Moreover, a protest meeting was held at Konnagar against septic tanks, which according to the *Bengalee* was attended by over 500 people, 'representing almost every family' of that area.[102] Thus, just before the publication of the Septic Tank Committee's interim report in December 1904, the atmosphere was considerably tensed.

After publication, the Indian press criticised the report from two main angles. First, the Committee was condemned for not ordering the immediate stoppage of septic tanks, even though their bacterial contents had now definitely been proven as potentially dangerous.[103] In this context, the Committee was also charged for its exclusive focus on the safety of the official municipal water supplies:

> There are lakhs of people living on both banks of the river, besides the luxurious inhabitants of Calcutta and Howrah, whose interests have to be considered. These people also have equal claims with the inhabitants of these two towns upon Government to have their health, purity and thirst attended to. [...] Let the Committee [...] consider how greatly the health of these

villages has suffered in the past form the discharge of 'raw sewage' into the river. Is not the existence of these latrines in connection with the mills one of the causes of the excessive prevalence of malaria in the villages about Calcutta?[104]

However, the Committee's conclusion that the purification of effluents was mandatory and its unequivocal support for their use as boiler feed-water gave rise to a cautious optimism, clearly linked to expectations that future policies would follow these lines.[105]

It was the Committee's treatment of the religious question that raised much stronger indignation. In a sharp commentary, the *Bengalee* hit straight at the Committee's fundamental argument, the distinction between (potentially valid) religious arguments and (allegedly irrelevant) sentimental motives:

> We find the Committee making the remark that the real objection on the part of the Hindu community to the pollution of the river appears to be sentimental rather than religious. We venture to think that this is a distinction without a difference. The sentimental objection has its roots in the deep-seated religious instinct of the people, which recoils with horror at the prospect of the pollution of the river, where rest the bones of their fathers and which is associated in their minds with ideas of eternal bliss.[106]

Moreover, the *Bengalee* challenged the exclusive claim for authority laid on the scriptures and the pundits:

> Text or no text, there the feeling is, deep-seated in the heart of the Hindoo who cannot cast his eyes upon the sacred stream without a feeling of deep reverence. It is easy enough to find texts on both sides of the question. But the official *Shastris* of the Government cannot, by any amount of casuistry, alter the testimony of living facts, or deny that the feeling against the pollution of the Bhagirathi, by transformed human excreta, is one of the strongest in the heart of the Hindoo. Let not our rulers quibble, or be led away by quibbling Pundits. Let them look at facts straight in the face and decide according to their testimony.[107]

The Committee's lack of consideration and understanding for popular feeling was generally attributed to its all-British membership. Thus, the *Bangavasi* noted:

> There is nothing strange in [its] opinion, considering that the Committee was composed solely of Europeans. The real feeling of a Hindu can be understood only by a Hindu. Even if the discharge of effluents is stopped at all points along the river above Tolly's Nullah, who can guarantee that this restriction will be enough to save the water of the *Adi Ganga* from pollution?[108]

Similarly, the *Hindoo Patriot* contended that the Committee, because of not containing 'a single Indian member', had 'deplorably' failed in consulting the 'religious feelings of the people', bowing to the mill managers' financial interests instead. On these grounds, it demanded the appointment of a new commission in which Indian opinion was adequately represented.[109] Several newspapers took this critique further, calling into question the British government's religious policy towards its Indian subjects. Thus, the *Hindoo Patriot* continued:

> [T]he step taken by Government in regard to the septic tank installation is calculated to create an impression in the public mind that it is indifferent to the traditions and religious sentiments of the people. [...] [T]he question is not how far, by the means of scientific method, the mills' effluent discharged into the Hooghly can be rendered innocuous, but whether pollution of the river is regarded as highly sacrilegious by the Hindus. Regarded in this light the report of the Committee which violates of the cardinal principles of British rule, viz., religious tolerance, should be received with the greatest caution.[110]

The *Hindoo Patriot*'s critique is reminiscent of an earlier one brought forward by the *Bengalee*, which accused government of failing to follow its own declared policy, 'not affording that religious protection to which the people are entitled'.[111] The *Howrah Hitaishi* more generally suggested that 'Englishmen nowadays show a tendency to belittle the Hindu religion'.[112]

Statements such as these are significant, since the Septic Tank Committee report came at a time when nationalist sentiment in Bengal was boiling high. From around the turn of the twentieth century, nationalist discourse had generally acquired a much sharper tone with the rise of the so-called extremist faction in- and outside the Indian National Congress, which demanded far-reaching changes in the political system and turned towards new forms of political mass agitation. Moreover, a growing number of small revolutionary groups promoted armed violence against the Raj.[113] Governor-General Lord Curzon's (1899–1905) conservative political agenda exacerbated tensions further. Curzon's proposal to split Bengal province for allegedly administrative reasons in 1903 met with fierce Indian opposition, but was also criticised by conservative British newspapers and the Bengal Chamber of Commerce. In the eyes of the Bengali nationalists, Curzon's move presented a thinly concealed attempt to curb their political influence. Consequently, hundreds of protest meetings were convened over a few months, and the annual sessions of the Indian National Congress in 1903 and 1904 passed resolutions condemning the proposal. In view of the potential threat of such a unified agitation, Curzon devised a more drastic, initially secret, scheme. He now envisaged the partition of Bengal into two separate provinces, one comprising Assam, North Bengal and East Bengal; the other West Bengal, Orissa and Bihar. Like this, the troublesome Bengali elites of Calcutta would be separated from those of the eastern districts, and both would constitute a minority in their respective provinces. By creating a Muslim majority province in the East, Curzon moreover hoped to obtain the support of the Bengali Muslims and to alienate them from the nationalist movement. Meanwhile, nationalist agitation continued, with hundreds of protest meetings and countless petitions launched over the months.[114] The Septic Tank Committee report, published in late December 1904, thus held considerable potential to exacerbate political conflict.

Considering the political climate one would expect that at least some nationalist figures sought to exploit the septic tank controversy for their political aims, even more so because the extremist faction had come to articulate their agenda around Hindu religious themes. With a view to turn Indian nationalism from an elite project into a mass movement, leaders such as Bal Gangadhar Tilak in Maharashtra and Aurobindo Ghose in Bengal employed a wide range of popular and emotionally appealing imaginaries which often carried very strong Hindu religious overtones. Tilak, for instance, converted the initially domestic Ganpati festival into a community-based affair in 1893, which included politically connoted processions, lectures and common worship.[115] Aurobindo Ghose and other Bengali extremists put forth their vision of the Indian nation as incarnation of the Hindu goddess Kali and, in line with other Hindu revivalists such as Swami Vivekananda, propagated the world-wide spread of Vedanta as the supreme religion. Hindu imageries were also extensively employed by small groups of Bengali revolutionaries (to which Aurobindo was closely related), who promoted violent resistance. Based on the ideal of selfless action put forth in the Bhagavad Gita, they conceived themselves as instruments in the hands of Kali, whose force was to bring about the destruction of colonial rule. While the first terrorist acts were carried out only in 1906, revolutionary cells such as the Maniktala group were active from 1902 onwards.[116] Hindu imageries were even more widely employed during the Swadeshi mass protests that erupted after the official declaration of partition in July 1905.[117]

However, there are no signs that Indian critiques against the Bengal government's septic tank policy were explicitly woven into the nationalist discourse to any significant extent. This is remarkable, given that nationalist leaders such as Bhupendra Nath Bose, Surendranath Banerjee and Narendra Nath Sen were prominently involved in the protests. Nor do the extremist and revolutionary nationalist factions seem to have taken the matter up, neither in 1904 nor in later years. The only hint towards the controversy's being put into an explicitly nationalist context is found in a British official's note of July 1904, which states that the provincial conference of the Indian National Congress at Burdwan had 'passed a resolution on the subject which emphasises its religious aspect'. The same official alleged that the controversy was engineered by Surendranath Banerjee. In his opinion, Banerjee wished 'to annoy the mill people, who succeeded some years ago in getting two new Municipalities formed' and with this greatly restricted Banerjee's 'range of [...] influence and patronage as Chairman of the North Barrackpore Municipality, and also to embarrass the Government'.[118] It would certainly be interesting to know whether or not (and if yes, to what extent) prominent political figures such as Banerjee, Sen and Bose, as well as the Indian press in general, were driven by political motives. However, this must remain open to speculation for the time being.

Even though the septic tank controversy was not directly incorporated into nationalist agitation, the delicate political context still had a direct influence on the Bengal government's subsequent river pollution policy. As we shall see, the intensity of Indian feeling on the septic tank issue and its potential for fuelling public protest caused the Bengal government to adopt a very careful, even secretive, approach after the publication of the Septic Tank Committee's report.

While the Septic Tank Committee treated Indian, and particularly Hindu religious opposition as a disturbing side-factor that needed to be neutralised some way or other, its stated objective was to find a technical solution to septic tank effluent disposal that guaranteed to safeguard public health. Both the Committee and other Bengal officials accepted the presence of pathogenic bacteria to be the major parameter for effluent and river water quality. However, many related details were very much open to debate. The discord between the sanitary board of Bengal and the health officer of Calcutta, Nield Cook, is exemplary for this. The sanitary board in its early defence of septic tanks had suggested the adoption of sand or cinder filters for effluent purification. Simultaneously, it denied the necessity of this, as the Massachusetts State Board of Health's research apparently proved that typhoid and *coli* bacteria got rapidly destroyed once they were exposed to sunlight. The special character of the Indian environment, the sanitary board explained – especially the Hooghly's large water volume and the 'tropical sun' acting on germs – crucially intensified river self-purification processes, and it was therefore only reasonable to let the Hooghly do its own share of the purification work.[119] While the sanitary board unanimously trusted the alleged rapid germicidal action taking place in river water, health officer Cook highlighted the lack of consensus among scientists about the efficacy of river self-purification. Referring to some American studies, Cook contended that pathogenic bacteria could travel very long distances in large rivers without losing any of their harmful potential. Based on this, he believed that even after a distance of 80 to 100 miles, absolute safety could not be guaranteed.[120] Thus, everybody agreed that diseases like typhoid and cholera were transmitted by waterborne germs, but it was not quite clear for how long these germs remained an actual threat once they had entered the Hooghly.

Neither the board nor Cook specified which studies exactly they based their opinions on. However, a look at the contemporary state of research on European and US American rivers conducted since the 1890s shows that the health officer's hesitations were well-founded. Generally, studies of that time maintained that river self-purification indeed eliminated germs and other organic, as well as inorganic, pollutants. However, the rate at which this process worked depended on the local characteristics of the factors involved, such as the amount of sewage, the river's rate of flow, the character of river beds and shores, and the amount of sediment a river carried. While the Danube was regenerated 25 miles below Vienna, the Isar below Munich still showed signs of pollution after 45 miles of flow.[121] For the Illinois river, the receptacle of Chicago's sewage, it took full

150 miles for the complete disappearance of *colon* bacilli.[122] The germicidal effect of sunlight – a principal argument in India in reference to the 'tropical' sun – was in fact controversially debated, different studies having furnished highly contradictory results as to the degree of its efficiency.[123] By the turn of the twentieth century, it was generally agreed that rivers owed most of their 'self-purification' to dilution and sedimentation, while other factors, such as sunlight exposure and aeration, were of secondary importance.[124] On these lines, Cook suggested that the mills may take advantage of the sunlight's agency by discharging the effluents during flood tide (so that germs would be exposed to it for a longer duration in the river's upper shallow reaches), but maintained that sunlight was only truly efficient in killing germs under laboratory conditions, where they could be spread out on a plate in a thin layer. In 'a large bulk of water' he insisted, 'sunlight has but little effect'.[125]

The Septic Tank Committee in principle sided with the sanitary board. The bacterial composition of the mills' septic tank effluents, the Committee acknowledged, was indeed highly problematic, the number of colon bacilli they contained being equal to that present in ordinary sewage. The results of its own research moreover corresponded with recent observations made by Alexander Cruikshank Houston at the London sewage outfall at Crossness and the British Royal Commission on Sewage Disposal, who both confirmed the problematic nature of sewage effluent gained through biological sewage treatment. According to Houston, pathogens such as the *B. Pyocyaneous* and the *B. Enteritidis Sporogenenes* survived biological sewage treatment through bacterial beds in large numbers, and he thought it unlikely that other pathogens, such as the typhoid bacillus, would not. Sewage treated by bacterial beds, Houston therefore concluded, could not be considered as any less harmful than original sewage. The British Royal Commission on Sewage Disposal presented similar conclusions. Reporting on septic tanks, the Royal Commission noted that the tanks did not succeed in digesting all the organic matter present in sewage, and that the effluent they produced was 'bacteriologically almost as impure as the sewage entering the tanks'.[126] The Septic Tank Committee however maintained that effluent quality itself could not be regarded as an isolated factor when assessing the potential danger of discharging effluents into the Hooghly. The degree of river pollution also depended on

[t]he rapidity with which further oxidation is produced and the bacteria are destroyed through contact with the river water and by the action of the air and the sun, and upon the volume of the water which passes down the river and with which the effluents are mixed.[127]

Therefore, any pollution caused by septic tank effluents in the Hooghly was likely to quickly disappear. Evidently, this conclusion contradicts the Committee's own investigations, which had revealed considerable bacterial contamination around the Palta water intake.[128] The Committee instead referred to a study on the sewage-polluted Barrow deep in Britain, which showed that the number of bacteria added to it by dumping tons of London sewage sludge declined rapidly.[129] Nevertheless, one significant problem remained: The Hooghly's low dry weather discharge in February and March. Even though investigations were still in process, the Committee had reason to believe that the latter stood at a mere 2,000 cubic feet per second (as against the earlier estimate of 20,000), a water volume similar to that of British rivers. During all other seasons, the Hooghly's large water volume 'in all probability' provided a sufficient safe-guard, neutralising problematic bacterial contents through dilution and self-purification, but during the dry months these powers were considerably diminished. Consequently, the Committee was forced to acknowledge that, under the present circumstances, the safety of the capital's municipal water supply was significantly endangered during the dry season. It therefore advised that the discharge of untreated septic tank effluents should be entirely stopped above the municipal water intakes, and be allowed only after Tolly's Nala, until a method was found to render them harmless.[130]

Overall, the Septic Tank Committee's approach perpetuated colonial perspec-tives on river pollution as established by the Government of India and the United Provinces in the 1890s. The guiding principle for effluent disposal, the Commit-tee maintained, was a river's assumed power of self-purification and dilution, which under the 'Indian paradigm' was believed to be greatly enhanced in Indian rivers. In the context of Calcutta however, this argumentative pattern created a stumbling block to the all-time indiscriminate discharge of sewage effluents (much as it did in Lucknow at around the same time),[131] due to the Hooghly's low water volume during the dry season. Other than in Banaras and Kanpur, where sewage and sullage could be discharged into a reasonably large river beyond city limits at all times, the Hooghly's low dry weather volume in combi-nation with the mills' location *above and within* the capital, put a serious chal-lenge before the Bengal government. The next and final section of this chapter shows how Bengal tried to overcome this obstacle.

The Septic Tank Committee's final report and its aftermath

The Septic Tank Committee report was, as mentioned previously, designated as an interim report, since two important investigations had not yet been concluded by the time of its publication in December 1904: The *definite* measurement of the Hooghly's water volume during the dry months – so crucial for an assess-ment of its self-purificatory powers during that season – and experiments into the efficacy of sand filtration for the further biological purification of septic tank effluents. By the end of 1905, all investigations had come to a close, and in January 1906, the Bengal government was ready to communicate its policy on the disposal of septic tank effluents.[132]

First, government noted that the Hooghly's minimum discharge during the dry months stood mostly above 3,500 cubic feet per second and rarely fell below 3,000 cubic feet. Therefore, measured by the river's powers of self-purification and dilution, pollution during the dry months was much less accentuated as initially apprehended by the Septic Tank Committee. Second, several methods had been put to test with regard to effluent purification. Experiments with sand filtration and copper sulphate had both furnished promising results; however, they were overall found to be impractical and too expensive. Sand filtration required the construction and maintenance of a very large filtration area, and it was doubtful whether either of the two methods could deal efficiently with effluents polluted to a higher degree than the present samples. Looking for an alternative, the Septic Tank Committee had then turned to the sterilisation of effluents with chlorinated lime. These experiments proved to be a great success. It was found that the addition of five grains[133] of chlorinated lime per each gallon of effluent rendered the effluent virtually sterile, and thus chemically and bacteriologically purer than the Hooghly water itself. Moreover, the method promised to be extremely cheap, amounting to around 10 to 15 rupees per month for an installation serving 2,000 people. Sewage treatment with chlorinated lime was a very recent innovation first applied in Britain and the US in the early 1890s, and was specifically aimed at the destruction of pathogenic bacteria. From the early twentieth century onwards, European countries and the US moreover turned to chlorination for the disinfection of their drinking water supplies.[134]

In view of the Septic Tank Committee's final report, the Bengal government reiterated its unfailing support for septic tanks. By operating tanks along the Committee's guidelines and by sterilising effluents with chlorinated lime, government declared, an effluent could be produced that was entirely free from objection from a chemical as well as bacteriological point of view. Thus, nothing stood in the way of existing and future septic tank installations if the mills complied with these conditions. However, special sanction from the Lieutenant-Governor of Bengal would be required for installing septic tanks near the Calcutta and Howrah water intakes.[135] The government's final verdict was received with consternation by the Indian press. Generally, it refrained from any further critique of how Hindu religion had been disregarded, resigning to the fact, as the *Bangavasi* put it, that 'the dicta of science are more worthy of respect than those of the Scriptures' to government. The proposed solutions to safeguard the river from bacteriological pollution, on the other hand, also failed to instil confidence. In the opinion of the *Bengalee*, the daily discharge of 1.5 million grains (i.e. almost 100 kilograms) of chlorinated lime into the Hooghly each day, floating up and down the river due to the tides, in itself presented a serious threat to aquatic life and public health.[136]

In order to assure the correct technical operation of septic tanks, the Bengal government called upon one of the most renowned British experts on sewage treatment technologies of the day, Gilbert J. Fowler.[137] Between February and April 1906, Fowler and the officiating sanitary commissioner of Bengal, W.W. Clemesha, conducted further inquiries into the pollution of the Hooghly by septic tank effluents.[138] Fowler was responsible for the chemical part of the investigation, while Clemesha took over the bacteriological analyses. The results were again quite sobering. While the chemical analyses proved overall satisfactory, the bacteriological analyses definitely confirmed existing concerns. Most of the effluents were 'far from good', Clemesha warned, and even the admission of relatively small amounts into the river created 'an extremely dangerous pollution' introducing a significant number of pathogenic bacteria into the latter. Again, samples taken near the water intake at Palta showed some pollution, which is why Clemesha suggested that water should be drawn from as far inside the stream as possible. Reporting on a sample taken at Kankinara, a town with five poorly working septic tanks, Clemesha came down heavily on advocates of river self-purification:

[I]t has been maintained by some people that the volume of water is so huge in the Hooghly that the small quantity of sewage that is put in by the towns on its banks cannot be recognized, nor does it pollute the water sufficiently to impair health. [...] Here we have a sample taken in a strong current a good way from the bank, a long way from any single mill outfall, with every condition favourable to rapid dilution, and yet considerable contamination can be demonstrated.[139]

The sanitary commissioner agreed that river self-purification did exist. But, he argued,

[t]he great fallacy of this line of reasoning is that Nature's processes, though complete, are very slow, and that long before complete oxidation can have taken place an untold amount of damage may have been done. [...] [I]t is not by any means yet certain that in a river the pathogenic bacilli disappear at the same rate as the organic matter is burnt up. [...] [E]ven if the river will eventually kill off all pathogenic bacilli that are put into it, the action is not so rapid as to make it impossible to trace pollution for miles below the point of discharge, and that therefore such pollution must be a constant menace to the health of the people using the water.[140]

In conclusion, Clemesha urged the sterilisation of the effluents by chloride of lime, which he believed would entirely remove pathogens from the septic tank effluents.

Clemesha's insistence on the treatment of septic tank effluents was shared by Gilbert J. Fowler. Based on his study of existing septic tank installations and experimental works at Kanchrapara and Entally, Fowler advised against the direct disposal of effluents into midstream and recommended their treatment either by aerobic methods (such as percolating filters or contact beds), or by sterilising them with chlorine. Moreover, Fowler gave extended advice on details of construction and operation, with which to render sewage treatment through septic tanks most effective.[141] Fowler's report including the results of the chemical and bacteriological analyses was ready for publication in 1906. However, subsequent events show just how sensitive this report was to the Bengal government. Fearing renewed agitation, Bengal requested the Government of India to defer the report's publication for the time being, a request which met with the latter's favour:

> The local Government's apprehension [...] seems very reasonable and the present is not the time at which to add to the Lieutenant-Governor's embarrassments. There is every chance that the publication of the report and the direction of attention to the septic tank system would at the present juncture give the Hindu agitators a new battle cry and we should be deafened with appeals to the populace to resist the desecration of the sacred river.[142]

In May 1907, Fowler wrote to the India Office, asking for permission to present his work on Calcutta septic tanks during the International Congress of Hygiene and Demography, held at Berlin in the same year. After being informed of the special circumstances surrounding the report, Fowler agreed to exclude from his presentation all reference 'to the special inquiry in Calcutta', limiting it to 'scientific and practical data and conclusions of general interest and importance'.[143] A look at his presentation, titled 'Report on the effect of the mechanical, chemical and biological purification of sewage. The treatment of sewage under tropical conditions' confirms that Fowler duly abided by this arrangement.[144] In August 1907, the report was printed in Britain, but its publication was still deferred.[145] In October 1908, the Government of India granted Bengal's request that the report should be reserved for the use of high government officials, and not be circulated generally.[146] As late as 1909, the Bengal government was anxious to keep Fowler's report secret. On the occasion of his visit to the United Provinces that year, Fowler asked for permission to visit the septic tank installations along the Hooghly. This was granted by the Bengal government, but only under the provision that his official report to the United Provinces did not contain any 'controversial matter affecting Bengal'.[147]

In the years following Fowler's report, the problems with septic tank effluent quality persisted. From 1906, the Bengal government undertook several administrative and legal steps to ensure that the mills duly chlorinated their septic tank effluents. In 1906, it included a section on the operation of septic tanks in the Bengal Factory Rules. First, inspectors of factories were to supervise septic tanks on a regular basis. Second, installations were to be kept clean and in working order, and the number of people using them was to be restricted in order to prevent overworking. Third, the discharge of septic tank effluents was not permitted unless it was 'clear, free from fœcal odour and non-putrescible [and had] been sterilized by the addition to each gallon of five grains of fresh chlorinated lime'. Commenting on the provisions, W.W. Clemesha urged that in view of the potential dangers involved for the city's water supply, any mill passing unsterilised effluents into the river should be immediately reported to the district magistrate.[148]

Implementation however proved difficult. In 1908, it was noted that many mills failed to properly observe the rules and that septic tank installations were often overcrowded and inefficiently operated. Consequently, the Lieutenant-Governor of Bengal announced his intention to appoint a special inspector of septic tank installations. At the same time, he proposed the mill managers the alternative of using the effluent as boiler feed-water. This practice, he maintained, was much simpler than effluent sterilisation and would evade the dangers inherent in passing 'an offensive and potentially dangerous liquid into the Hooghly'. If the mill managers declined this alternative, the Lieutenant-Governor warned, government would be forced to appoint a special inspector of septic tank installations and take 'much more stringent measures than hitherto' to ensure the safety of public health.[149] The mill managers, however, were not in favour of the Lieutenant-Governor's proposal. In a letter to government, the Bengal Chamber of Commerce pointed out that mill managers were still apprehensive of using septic tank effluents as boiler feed-water, convinced that 'caste prejudices' would prevent their workmen from handling boilers in which effluents were used. Thus, fearing 'endless difficulties' to arise from trying to implement the practice, they expressed their preference towards the appointment of a special inspector, given he be a 'competent and reliable European' from a 'desirable class of men'.[150] Consequently, the Bengal government appointed J. Dallas, 'a Scotchman with a good knowledge of the Hooghly' and a 'thorough knowledge of the working of septic tanks', for the post.[151] Dallas had previously worked for the Standard Jute Mills, and the cordial relations he maintained with the mill managers from that time, the sanitary commissioner remarked, would contribute to his ability of dealing with them.[152]

In 1910, W.W. Clemesha issued a follow-up report to his and Fowler's earlier reports of 1906, presenting an inquiry into the working of existing septic tank installations in the mills.[153] Overall, the verdict was negative. The majority of the installations, Clemesha wrote, were not given adequate attention. Regular complaints received pertained to 'flushes being out of order, deficiency of water-supply [and] lack of attention to the cleaning of filters', and many tanks were overworked. Moreover, the chlorination of effluents, so crucial to ensure the neutralisation of harmful bacterial contents, was done satisfactorily only by very few mills. The present state of things, Clemesha remarked, did 'not reflect any very great credit on the mill owners as a whole'.[154]

Clemesha's report did not include any bacteriological analyses of river water. However, at around the same time, the Calcutta municipal board sat over a report submitted by its health officer, who had examined the river water as well as the filtered water supply at Palta. While the filters at Palta appeared to work very well, furnishing a drinking water of good quality, the numbers of bacteria in the ordinary river water was found to be so disturbingly high that some municipal commissioners doubted their accuracy. The health officer conceded that bacteriological water analysis was indeed a complex process, liable to errors, but maintained that the figures were correct and that earlier samples had shown even higher figures.[155] Judged in the light of Clemesha's findings, and even more so in the light of subsequent developments, it seems well justified to conclude that the Bengal government by 1910 still failed to properly enforce the sterilisation of effluents, and that the Hooghly was still being severely polluted.

Notes

1 At the head of the Bengal delta near Farakka the Ganges splits up into two major channels, the Padma and the Bhagirathi. The Padma continues its flow eastwards into Bangladesh, while the Bhagirathi takes a sharp turn to the South, passing by Kolkata and emptying into the sea some 130 kilometres below the city. The Bhagirathi is commonly held to be the original Ganges as it used to carry the main water flow until the late sixteenth century, when the latter suddenly shifted along the course of the Padma due to natural causes. Since British rule, the Bhagirathi is generally referred to as Hooghly, called after a major trading centre of the same name that existed long before the British set foot into the region (Das Gupta 1990: 2; see also Thomas 2004: 11–14).

2 Goode 1916: 361.

3 Sinha 1990: 32–3; Sau 1997: 14, 36–40. On the opposite side of the Hooghly, an urban settlement of similar size arose, growing into what is today known as Haora, referred to as 'Howrah' during the nineteenth and early twentieth centuries (Nair 1990: 13).

4 Datta 2009.

5 Nath and Majumdar 1990: 167; Biswas 1990: 160.

6 Furedy 1987.

7 See also Chakrabarti 2015: 184–91.

8 Goode 1916: 168–70.

9 See, for example: *ARSC Bengal for the year 1879*, p. 82; *ARSC Bengal for the year 1886*, p. 91.

10 Silk 1903: 1, 26–7.

11 Captain J. Mulvany, Superintendent of the Presidency Jail, to Inspector-General of Jails, Bengal, 29.8.1902, APAC, IOR, P/6564, GoBeng, Municipal Dpt., Municipal Branch, Prgs April 1903, Nos 47–8.

12 H.S. McGill, Special Sanitary Officer, Army Head-Quarters, India, 'Report on septic tank installation at Fort William, Calcutta', APAC, IOR, P/6564, GoBeng, Municipal Dpt., Municipal Branch, Prgs April 1903, No. 53.

13 E.W. Collin, Offg. Secy GoBeng, Municipal Dpt., to Secy, Sanitary Board Bengal, 11.3.1903, APAC, IOR, P/6564, GoBeng, Municipal Dpt., Municipal Branch, Prgs April 1903, No. 49.

14 L.P. Shirres, Offg. Secy GoBeng, Municipal Dpt., to Secy GOI, Home Dpt., 29.5.1903, NAI, GOI, Home Dpt., Sanitary Branch, Prgs June 1903, No. 295.

15 'Proceedings of the Sanitary Board, Bengal', 5.3.1903, APAC, IOR, P/6564, GoBeng, Municipal Dpt., Municipal Branch, Prgs April 1903, No. 51; A.E. Silk, Secy Sanitary Board Bengal, to Secy GoBeng, Municipal Dpt., 4.4.1903, APAC, IOR, P/6564, GoBeng, Municipal Dpt., Municipal Branch, Prgs April 1903, No. 57.

16 Sethia 1996; Rothermund 1988: 56–60.

17 Chakrabarty 1989: 7–8.

18 C.A. Walsh, Special Inspector of Factories, to Secy GoBeng, Municipal Dpt., 18.12.1894, APAC, IOR, P/4738, GoBeng, Municipal Dpt., Municipal Branch, Prgs March 1895, No. 27.

19 Chakrabarty 1989: 10.

20 Note by B.H. Deare, Sanitary Commissioner Bengal, and A.E. Silk, Sanitary Engineer Bengal, 9.1.1903, NAI, GOI, Home Dpt., Sanitary Branch, Prgs. June 1903, No. 295.

21 *Bengalee*, 6 September 1903, p. 3.

22 *Bengalee*, 17 September 1903, p. 3.

23 APAC, IOR, P/6565, GoBeng, Municipal Dpt., Municipal Branch, Prgs December 1903, Nos 30–1.

24 Ibid.

25 F.C. Clarkson, Sanitary Commissioner Bengal, to Secy GoBeng, Municipal Dpt., 14.10.1903, APAC, IOR, P/6796, GoBeng, Municipal Dpt., Municipal Branch, Prgs February 1904, No. 22.

26 'Proceedings of the Sanitary Board, Bengal', 5.3.1903, APAC, IOR, P/6564, GoBeng, Municipal Dpt., Municipal Branch, Prgs April 1903, No. 51; A.E. Silk, Secy Sanitary Board Bengal, to Secy GoBeng, Municipal Dpt., 4.4.1903, APAC, IOR, P/6564, GoBeng, Municipal Dpt., Municipal Branch, Prgs April 1903, No. 57.

27 APAC, IOR, L/R/5/29, *Bengal Newspaper Reports [henceforth: BNR], 1903*, 'Sri Sri Vishnu Priya-o-Ananda Bazar Patrika', 11 November 1903.

28 APAC, IOR, L/R/5/29, *BNR, 1903*, 'Sanjivani', 3 December 1903.

29 *Bengalee*, 11 December 1903, p. 3.

30 *Bengalee*, 13 January 1904, p. 3.

31 APAC, IOR, P/6796, GoBeng, Municipal Dpt. Municipal Branch, Prgs February 1904, No. 24.

32 Ibid.; APAC, IOR, L/R/5/29, *BNR, 1903*, 'Basumati', 19 December 1903.

33 APAC, IOR, L/R/5/29, *BNR, 1903*, 'Sanjivani', 3 December 1903; *Bengalee*, 11 December 1903, p. 3. See also APAC, IOR, L/R/5/29, *BNR, 1903*, 'Basumati', 19 December 1903.

34 APAC, IOR, P/6796, GoBeng, Municipal Dpt. Municipal Branch, Prgs February 1904, No. 25.

35 L.P. Shirres, Secy GoBeng, Municipal Dpt., to Secy Sanitary Board Bengal, 9.2.1904, APAC, IOR, P/6796, GoBeng, Municipal Dpt. Municipal Branch, Prgs February 1904, No. 30.

36 A.E. Silk, Secy Sanitary Board Bengal, to Secy GoBeng, Municipal Dpt., 22.3.1904, APAC, IOR, P/6797, GoBeng, Municipal Dpt., Municipal Branch, Prgs August 1904, No. 10.

37 Ibid.

38 Proceedings of the Sanitary Board, Bengal, 26.2.1904, APAC, IOR, P/6797, GoBeng, Municipal Dpt., Municipal Branch, Prgs August 1904, No. 11.

39 Silk 1903: 24–5.

40 A.E. Silk, Secy Sanitary Board, Bengal, to Secy GoBeng, Municipal Dpt., 22.3.1904, APAC, IOR, P/6797, GoBeng, Municipal Dpt., Municipal Branch, Prgs August 1904, No. 10.

41 *Bengalee*, 3 April 1904, p. 3.

42 Bhupendra Nath Bose [also: Basu] (1859–1924) counted among India's foremost nationalist leaders. A long-standing member of the Indian National Congress and member of the Bengal Legislative Council from 1904 to 1910, Bose was actively involved in the Swadeshi movement and in 1905 presided over the Bengal Provincial Conference held at Mymensingh. In 1914, he was elected as president of the Indian National Congress (Hayavadana Rao [1916]; Sarkar 1973: 117, 221–2). Surendranath Banerjee (1848–1925) was equally a key figure of Indian nationalism. In 1876 he founded the Indian Association, one of the earliest Indian political organisations, which spearheaded agitations such as those against the Arms Act (1877) and the Vernacular Press Act (1878). He was involved with a wide range of political and social organisations, such as the Calcutta Corporation (1876–99) and the Bengal Legislative Council (1893–1901), and in 1894 and 1902 officiated as president of the Indian National Congress. Just like Bhupendra Nath Bose, Banerjee belonged to the moderate faction of the Indian nationalist movement and was actively involved in the Swadeshi protests. In 1879, Banerjee took over the *Bengalee*, which he edited for nearly 40 years from a nationalist standpoint (Sinha 1968b).

43 APAC, IOR, P/6797, GoBeng, Municipal Dpt., Municipal Branch, Prgs August 1904, Nos 10–11.

44 APAC, IOR, P/6797, GoBeng, Municipal Dpt., Municipal Branch, Prgs August 1904, No. 12.

45 *Report of the Special Committee Appointed to Examine the Working of the Septic-Tank Installations in Bengal*, Calcutta: Bengal Secretariat Press, 1905 [hereafter: Septic Tank Committee Report].

46 Ibid., pp. 1–2.

47 Ibid., pp. 2–3 and Appendix, pp. viii–xii. Looking at the reports submitted by the bacteriological examiners, this last statement seems quite attenuated. L. Rogers reported 'a high degree of bacterial pollution' around the Palta waterworks (ibid., Appendix, p. xii), while N. Cook thought the pollution levels constituted a possible danger to municipal water supplies (ibid., Appendix, pp. xxii–xxiv, esp. xxiii).

48 Ibid., p. 4.

49 Ibid., pp. 4–6.

50 Ibid., pp. 6–10.

51 Raja Peary Mohun Mookerjee was a wealthy aristocrat and landowner, and the 'moving spirit' behind the British Indian Association (Trevithick 2006: 141, 159). Baman Das Banerjee, according to an official report, was a municipal commissioner and honorary magistrate at Serampore, and one of the 'few respectable residents' of Rishra (T. Inglis, Offg. Commissioner, Burdwan Division, to Secy GoBeng, Municipal Dpt., 25.1.1904, APAC, IOR, P/6796, GoBeng, Municipal Dpt., Municipal Branch, Prgs February 1904, No. 27). Narendra Nath Sen was actively involved in the Indian nationalist movement, closely associated with both the British Indian Association and the Indian Association, and took on a leading role during the Swadeshi movement. As the editor and proprietor of the *Indian Mirror*, Sen moreover consistently fought for the freedom of the press (Sinha 1968a: 292–5). Nalin Bihari Sircar was a business man and municipal commissioner (Mukharji 2011: 154; Mukherjee 1968: 106). On Abinash Chandra Banerjee, no information could be obtained. For the sake of consistency, the witnesses' names here and in what follows are spelt the way they appear in the Septic Tank Committee's report.

52 Testimony of Narendra Nath Sen, *Septic Tank Committee Report*, Appendix, pp. xxxviii–xxxix.

53 Testimony of Narendra Nath Sen, ibid., p. xxxix, and testimony of Baman Das Banerjee, ibid., p. xvii (Banerjee's testimony essentially reproduces the contents of the Rishra petition).

54 Testimony of Abinash Chandra Banerjee, ibid., p. xix. See also the testimony of Raja Peary Mohun Mookerjee, ibid., p. xv, and the testimony of Babu Baman Das Banerjee, ibid., pp. xvi–xvii.

55 K. Goswami, Chairman, Serampore Municipality, to the Magistrate of Hooghly, 22.12.1903, APAC, IOR, P/6796, GoBeng, Municipal Dpt., Municipal Branch, Prgs February 1904, No. 28.

56 Wohl 1983: 234.

57 Turner 1981.

58 APAC, IOR, L/R/5/30, *BNR, 1904*, 'Howrah Hitaishi', 20 February 1904.

59 Testimony of Nalin Bihari Sircar, *Septic Tank Committee Report*, Appendix, p. xviii.

60 Note by W.J. Simpson, n.d., APAC, IOR, P/4738, GoBeng, Municipal Dpt., Municipal Branch, Prgs March 1895, No. 24.

61 J.A. Bourdillon, Offg. Secy GoBeng, to Secy Bengal Chamber of Commerce, 15.12.1894, APAC, IOR, P/4738, GoBeng, Municipal Dpt., Municipal Branch, Prgs March 1895, No. 25.

62 S.E.J. Clarke, Secy Bengal Chamber of Commerce, to Offg. Secy GoBeng, 8.1.1895, APAC, IOR, P/4738, GoBeng, Municipal Dpt., Municipal Branch, Prgs March 1895, No. 28.

63 C.A. Walsh, Special Inspector of Factories, to Secy GoBeng, Municipal Dpt., 24.12.1894, APAC, IOR, P/4738, GoBeng, Municipal Dpt., Municipal Branch, Prgs March 1895, No. 27.

64 Chakrabarty 1989: 74–6, 80–1, 168–9.
65 See the board's statement quoted earlier, supporting the mill managers' contention that it would be inappropriate to force additional expenditure on them for the 'ultra-purification of the filtrates [...] while the pollution of the Hooghly from other sources is so patent' (A.E. Silk, Secy Sanitary Board, Bengal, to Secy GoBeng, Municipal Dpt., 22.3.1904, APAC, IOR, P/6797, GoBeng, Municipal Dpt., Municipal Branch, Prgs August 1904, No. 10; H.C. Woodman, Under-Secy GoBeng, to Sanitary Board, Bengal, 9.2.1904, APAC, IOR, P/6797, GoBeng, Municipal Dpt., Municipal Branch, Prgs August 1904, No. 10). Elsewhere, commenting on the assumed success of the septic tank at Presidency Jail, the board noted that septic tanks had proven to be 'of great use to the commercial community' ('Extract from the proceedings of the Sanitary Board, Bengal', n.d., APAC, IOR, P/6564, GoBeng, Municipal Dpt., Municipal Branch, Prgs April 1903, No. 55).
66 Testimony of Nalin Bihari Sircar, *Septic Tank Committee Report*, Appendix, p. xviii. Reporting on the same matter earlier, the *Bangavasi* blamed the polluted condition of the Bally creek on 'the large quantity of refuse matter (both solid and liquid) which the mills are allowed to discharge on its banks', and reminded its readers that even though reforms had been promised in 1895, nothing had changed since then (APAC, IOR, L/R/5/30, *BNR, 1904*, 'Bangavasi', 6 August 1904).
67 Testimony of Nalin Bihari Sircar, *Septic Tank Committee Report*, Appendix, p. xix.
68 Testimony of Narendra Nath Sen, ibid., p. xxxix.
69 Testimony of Raja Peary Mohun Mookerjee, ibid., p. xvi.
70 Testimony of Nalin Bihari Sircar, ibid., p. xix. Sircar refers to the Dharmashastras, a Sanskrit text corpus containing religious and legal injunctions. The Manu Samhita (more commonly called Manusmriti) sets out rules of conduct and obligations for men, women, wives, different castes, different stages of life, etc. From the second century CE onwards, writers have looked at it as the most authoritative text among the Dharmashastras (Brockington 2008: 186–8; Schmiedchen 2008: 491).
71 Testimony of Baman Das Banerjee, *Septic Tank Committee Report*, Appendix, p. xvii.
72 Testimony of Baman Das Banerjee, ibid., p. xviii.
73 Testimony of Abinash Chandra Banerjee, ibid., p. xx.
74 Testimony of Raja Peary Mohun Mookerjee, ibid., p. xvi.
75 See also testimony of Raja Peary Mohun Mookerjee, ibid.
76 Testimony of Yogisa Chandra Sastree, ibid., p. xxv. On Sastree see also Chakrabarti 2015: 201–3.

77   In *Brahmandapuranam* it has been stated that one must abandon the undermentioned thirteen acts after reaching Ganges water: – (1) Cleansing (after the evacuation of excrement). (2) Washing of the face (after a meal). (3) Sprinkling (of other water) [which] may also be translated by 'evacuation of excrement' and 'making water,' respectively. (4) Throwing in flowers, etc. (previously used for worshipping any other deity). (5) Rubbing off the dirts (of body). (6) Shampooing. (7) Sports (in water of Ganges). (8) Receiving a donation. (9) Showing the want of faith towards Ganga. (10) Showing attachment to any other pilgrimage. (11) Eulogizing of any other pilgrimage. (12) Changing the clothes and washing of them. (13) Swimming.

(Note by Yogisa Chandra Sastree, *Septic Tank Committee Report*, Appendix, p. xxxiv)

78 Ibid. In his note, Sastree distinguishes between material and ritual pollution by writing 'dirty' water when referring to material pollution (for the Sanskrit *anirmala jal* and *apariskrit jal*) and writing 'defiled', 'polluted' or 'unholy' when referring to ritual pollution (for the Sanskrit *apavitra jal*), see ibid., p. xxxii.
79 The respective verses in the Manu Samhita are found in Discourse 4, Verse 56 ('He shall not throw into water urine, or faeces, or spittings, or anything else

contaminated by unclean things, or blood or poisons') and Discourse 5, Verse 107 ('What needs purification is purified by clay and water; the river is purified by its current; the woman of unclean mind by menstruation; and Brahmanas by renunciation'). See Jha (1920–6).

80 I was unable to locate this passage in the Vishnu Purana. However, Book 3, Chapter 11 contains the injunction that

> A wise man will never [...] pass either excrement in a ploughed field, or pasturage, or in the company of men, or on a high road, or in rivers and the like, which are holy, or on the bank of a stream [...].

> (Wilson 2001)

81 This view, one may note, presents an interesting intersection between secular and spiritual concepts of river self-purification. According to Hindu religion, motion also essentially determines the purifying powers of water itself. While it is held that all waters possess the power to effect ritual purification, the latter is believed to be greatly enhanced in moving or flowing water (Kinsley 2005: 189–91; Eck 1993: 217).

82 Note by Yogisa Chandra Sastree, *Septic Tank Committee Report*, Appendix, p. xxxv.

83 Testimony of Satis Chandra Vidyabhusana, ibid., p. xliv; testimony of Kaliprosanna Bhattacharya, ibid., p. xli; testimony of Rajani Kanto Vidyaratna, ibid., p. xlviii.

84 Testimony of Satis Chandra Vidyabhusana, p. xliv; testimony of Raj Krishna Tarkapanchanan, ibid., p. xlvii.

85 Note by Kaliprosanna Bhattacharya, ibid., p. xlii (emphasis in the original).

86 Testimony of Raj Krishna Tarkapanchanan, ibid., p. xlvii.

87 Testimony of Raja Peary Mohun Mookerjee, ibid., p. xv.

88 Testimony of Yogisa Chandra Sastree, ibid., p. xxv.

89 Testimony of Satis Chandra Vidyabhusana, ibid., p. xliv.

90 Testimony of Raj Krishna Tarkapanchanan, ibid., p. xlvii.

91 *Septic Tank Committee Report*, p. 4.

92 Hatcher 2004: 182–4; Chakrabarti 2004: 11. On Indian socio-religious reform movements and their leaders during the nineteenth and early twentieth centuries see Baird (2001).

93 Raina and Habib 1995: 95. In fact, they counted several outstanding scientists among their own, such as chemist Prafulla Chandra Ray, geologist Pramatha Nath Bose and physicist Jagadis Chandra Bose. However, these men faced intense racial prejudice and discrimination on part of the British (who claimed Bengalis to be unfit to conduct original scientific research), which is why only few of them got recruited to significant posts within the scientific services (Arnold 2000: 139–41).

94 *Bengalee*, 24 August 1904, p. 3.

95 Ibid.

96 See, for example, APAC, IOR, L/R/5/30, *BNR, 1904*, 'Sri Sri Vishnu Priya-o-Ananda Bazar Patrika', 23 March 1904, 'Amrit Bazar Patrika', 11 June 1904, and 'Sanjivani', 16 June 1904; as well as *Bengalee*, 8 October 1904, p. 3.

97 *Bengalee*, 13 August 1904, p. 3.

98 Pitra Paksh is an important Hindu festival taking place during the dark fortnight of the Hindu month Ashvina (September-October). During Pitra Paksh, Hindus offer daily sacrifices called *śrāddha*, consisting of water and different food items, such as rice balls, to their ancestors. The ritual is deemed essential for the peaceful afterlife of the ancestors, which in turn promote the interests of their living descendants (Seghal 1999: 505–6; Nelson 2007: 1279).

99 *Bengalee*, 2 October 1904, p. 3; *Bengalee*, 8 October 1904, p. 3 (quote ibid.).

100 APAC, IOR, L/R/5/30, *BNR, 1904*, 'Sri Sri Vishnu Priya-o-Ananda Bazar Patrika', 20 January 1904; *Bengalee*, 6 September 1903, p. 3. See also *Bengalee*, 30 April 1904, p. 3.

101 APAC, IOR, L/R/5/30, *BNR, 1904*, 'Sanjivani', 16 June 1904 and 'Bangavasi', 11 June 1904.
102 *Bengalee*, 17 December 1904, p. 3.
103 APAC, IOR, L/R/5/31, *BNR, 1905*, 'Amrita Bazar Patrika', 23 January 1905; *Bengalee*, 11 July 1905, p. 3.
104 APAC, IOR, L/R/5/31, *BNR, 1905*, 'Samvad Prabhakar', 10 January 1905; see also ibid., 'Reis and Rayyet', 28 January 1905.
105 *Bengalee*, 18 January 1905, p. 3; APAC, IOR, L/R/5/31, *BNR, 1905*, 'Indian Mirror', 16 January 1905 and 'Hindoo Patriot', 20./23 January 1905.
106 *Bengalee*, 18 January 1905, p. 3.
107 Ibid.
108 APAC, IOR, L/R/5/31, *BNR, 1905*, 'Bangavasi', 14 January 1905. See also APAC, IOR, L/R/5/30, *BNR, 1904*, 'Sri Sri Vishnu Priya-o-Ananda Patrika', 17 August 1904.
109 APAC, IOR, L/R/5/31, *BNR, 1905*, 'Hindoo Patriot' 20./23 January 1905.
110 Ibid., 'Hindoo Patriot', 18 January 1905.
111 *Bengalee*, 8 October 1904, p. 3.
112 APAC, IOR, L/R/5/31, *BNR, 1905*, 'Howrah Hitaishi', 21 January 1905.
113 Ray 1988: 62–89; Heehs 1998.
114 Majumdar 1981: 18–68.
115 Tejani 2008: 53–61; Nejad 2015: 273.
116 Boehmer 2002; Gupta 1997: 3–27; Heehs 2010: 153–76. See also Heehs 1998: 96–123.
117 Sarkar 1973: 303–16.
118 H.H. Risley, 4.7.1904, NAI, GOI, Home Dpt., Sanitary Branch, Prgs September 1904, Nos 158–9, Notes.
119 A.E. Silk, Secy Sanitary Board, Bengal, to Secy GoBeng, Municipal Dpt., 22.3.1904, APAC, IOR, P/6797, GoBeng, Municipal Dpt., Municipal Branch, Prgs August 1904, No. 10.
120 'Proceedings of the Sanitary Board, Bengal', 26.2.1904, APAC, IOR, P/6797, GoBeng, Municipal Dpt., Municipal Branch, Prgs August 1904, No. 11.
121 Turneaure and Russell 1901: 149–56.
122 Jordan 1903: 89.
123 Turneaure and Russell 1901: 149–56.
124 Ibid.: 155; Sedgwick 1902: 127–9, 132–4, 231–3; Jordan 1903.
125 *Septic Tank Committee Report*, Appendix, p. xxiii.
126 *Septic Tank Committee Report*, pp. 2–3; Royal Commission on Sewage Disposal 1915: 5. As the Royal Commission's fifth report dealing with septic tanks was published only in 1905, it appears that the Indian Committee received the respective information beforehand through another channel. For a detailed account on the experiments made with biological sewage treatment at the London sewage outfalls during the 1890s and early 1900s, see Clowes and Houston (1904).
127 *Septic Tank Committee Report*, p. 3.
128 *Septic Tank Committee Report*, Appendix, pp. xii and xxii–xxiv.
129 *Septic Tank Committee Report*, p. 3. From 1887 onwards, the sludge gained by treating London's sewage at Crossness and Barking was transported away by sludge boats and dumped into the North Sea at different sites. This practice was stopped only in 1998, when Britain complied with European Union legislation against marine waste disposal. Since then, the sludge is burnt in two incineration plants right at Crossness and Barking, and is used for power generation (Halliday 2001: 106–7).
130 *Septic Tank Committee Report*, p. 4.
131 As it will be remembered, the United Provinces considered the discharge of untreated sewage into the Gomti at Lucknow dangerous all over the year due to that

river's small water volume, and the city ultimately retained the hand-removal system for excreta and built sullage farms for the remaining wastewater (see Chapter 4).

132 Resolution No. 418, 6.1.1906, APAC, IOR, P/7303, GoBeng, Municipal Dpt., Sanitation Branch, Prgs February 1906, No. 3.

133 One grain is equal to 64.79891 milligrams.

134 Race 1918: 7–12; Jones 2013: 105–21. Joseph Race in his publication lists the Septic Tank Committee's research on chlorination among several important works undertaken in this context, an indication that the Committee's work was received by the scientific community beyond British India. Race also mentions the work of an IMS officer called Vincent B. Nesfield a few years earlier. Nesfield experimented with different chlorine compounds for the destruction of pathogens in water, looking for a method that could be easily used on field during military campaigns (Race 1918: 7–8).

135 Resolution No. 418, 6.1.1906, APAC, IOR, P/7303, GoBeng, Municipal Dpt., Sanitation Branch, Prgs February 1906, No. 3.

136 *Bengalee*, 2 March 1906, p. 3. The calculation is based on an average of 5 grains per 1 gallon of effluent, with an assumed daily total of 300,000 gallons of effluents. The *Bengalee* did not stand alone with its concerns over possible adverse effects of chlorination on water quality. In Europe and the US, the chlorination of water supplies was repeatedly criticised for a variety of issues. Amongst others, it was believed that chlorinated water caused colic, killed fish and birds, and destroyed plants and flowers (Race 1918: 62–71).

137 On Fowler see Chapter 4.

138 W.W. Clemesha, 'Note on the bacteriological and chemical examination of the Hooghly water', 10.5.1906, APAC, IOR, P/7326, GOI, Home Dpt., Sanitary Branch, Prgs October 1906, No. 293.

139 Ibid.

140 Ibid.

141 'Sanitary Commissioner's note on Dr. Fowler's report on the purification of sewage in connection specially with the Jute Mills on the Hooghly', APAC, IOR, P/7864, GoBeng, Municipal Dpt., Sanitation Branch, November 1908, No. 18.

142 J.C. Ferguson, Under Secy, GOI, Home Dpt., 7.5.1907, NAI, GOI, Home Dpt., Sanitary Branch, Prgs May 1907, No. 294.

143 APAC, IOR, P/7602, GOI, Home Dpt., Sanitary Branch, July 1907, No. 182.

144 Fowler 1908: 46–63.

145 APAC, IOR, P/7602, GOI, Home Dpt., Sanitary Branch, September 1907, No. 303.

146 NAI, GOI, Home Dpt., Sanitary Branch, November 1908, Nos 25–7.

147 APAC, IOR, P/8140, GoBeng, Municipal Dpt., Sanitation Branch, Prgs April 1909 (B), Nos 39–42.

148 T.W. Richardson, Offg. Secy GoBeng, 'Notification', APAC, IOR, P/7601, GOI, Home Dpt., Sanitary Branch, Prgs January 1907, No. 39; W.W. Clemesha, Offg. Sanitary Commissioner Bengal, to the Inspector of Factories, 23.6.1906, APAC, IOR, P/7601, GOI, Home Dpt., Sanitary Branch, Prgs January 1907, No. 39.

149 C.E.A.W. Oldham, Secy GoBeng, Municipal Dpt., to Secy Bengal Chamber of Commerce, 17.10.1908, APAC, IOR, P/7864, GoBeng, Municipal Dpt., Sanitary Branch, Prgs November 1908, No. 23.

150 H.M. Haywood, Secy Bengal Chamber of Commerce, to Secy GoBeng, Municipal Dpt., 13.1.1909, APAC, IOR, P/8140, GoBeng, Municipal Dpt., Sanitation Branch, Prgs April 1909, No. 3; H.M. Haywood, Secy Bengal Chamber of Commerce, to Secy GoBeng, Municipal Dpt., 20.3.1909, APAC, IOR, P/8140, GoBeng, Municipal Dpt., Sanitation Branch, Prgs April 1909, No. 7.

151 C.E.A.W. Oldham, Secy GoBeng, Municipal Dpt., to Secy Bengal Chamber of Commerce, 6.4.1909, APAC, IOR, P/8140, GoBeng, Municipal Dpt., Sanitation Branch, Prgs April 1909, No. 11.

152 J. Dallas, Inspector of Septic Tanks, to Sanitary Commissioner Bengal, 16.7.1909, APAC, IOR, P/8142, GoBeng, Municipal Dpt., Sanitation Branch, Prgs September 1909, No. 12; H. Wheeler, Secy GoBeng, Municipal Dpt., to Secy GoBeng, Home Dpt., 22.9.1909, APAC, IOR, P/8142, GoBeng, Municipal Dpt., Sanitation Branch, Prgs September 1909, No. 13.

153 After Fowler's departure from Bengal in 1906, Clemesha had continued his investigations into septic tanks, and kept taking analyses of river water and tank effluents. For this purpose, the Bengal government sanctioned the establishment of a special laboratory attached to the Sanitary Department (W.W. Clemesha, Offg. Sanitary Commissioner Bengal, to Secy GoBeng, Municipal Dpt., 24.8.1906, and H.J. McIntosh, Offg. Secy GoBeng, Municipal Dpt., to W.W. Clemesha, Offg. Sanitary Commissioner Bengal, 13.10.1906, NAI, GOI, Home Dpt., Sanitary Branch, Prgs May 1908, Nos 29–30.

154 W.W. Clemesha, 'Report on septic tanks in Bengal', n.d., APAC, IOR, P/8418, GoBeng, Municipal Dpt., Sanitation Branch, Prgs February 1910, Nos 16–17.

155 C.F. Payne, Acting Chairman, Corporation of Calcutta, to Secy GoBeng, Municipal Dpt., 27.9.1909, APAC, IOR, P/8142, GoBeng, Municipal Dpt., Municipal Branch, Prgs December 1909, Nos 35–6.

Conclusion

This book has offered a first in-depth analysis of early colonial river pollution policy in India, a topic so far largely unexplored by scholars writing on South Asia's environmental history. To be sure, a single book cannot attempt a comprehensive treatment of the subject matter, and it is to be hoped that future studies will fill the many gaps that remain. With that in mind, this conclusion aims not only to connect the main findings of the preceding chapters, but also to highlight certain areas that may constitute fertile grounds for further research.

A basic contention of this study, set out in the introduction, is that the roots of independent India's enduring river pollution crisis lie with British colonial policies on wastewater disposal devised during the late nineteenth and early twentieth centuries. A look at events in Calcutta and other riparian cities during the closing decades of colonial rule makes this evident, revealing, as it does, that the United Provinces and Bengal both kept perpetuating the policy patterns established around the turn of the twentieth century. In the United Provinces, colonial fiscal conservatism crystallised as the major guiding principle in directing sewerage infrastructure development. Consequently, existing sewerage systems and sewage treatment facilities fell rapidly into decay, and with municipalities possessing neither sufficient funds nor trained staff to remedy this situation, river pollution was on a steady increase. In Bengal, the state government stuck to its industry-friendly policy in Calcutta, even while knowing about the intense and growing pollution caused by the mills' septic tank effluents. By the 1930s then, the colonial legacy of river pollution which independent India was to inherit was already clearly manifest.

In Bengal, various official reports produced during the 1910s and 1920s showed that the Hooghly at Calcutta was being contaminated with faecal matter to an alarming extent. In 1918, Bengal's director of public health[1] issued a warning to all riparian municipalities that unless Hooghly water was boiled or filtered, it was safe neither for consumption nor for cleansing dishes and utensils. For the majority of Indians in the Bengal Legislative Council it was plainly obvious that septic tank effluents were the major source of faecal contamination, and the controversy that had lain dormant for some years flared up again. In a resolution aiming at tighter state control over the mills, Indian council members demanded that the septic tank inspector be directed to personally ascertain the quality of septic tank effluents before each release into the river, instead of inspecting the installations only every now and then. The Bengal government vehemently rejected this suggestion. Essentially, it agreed that the Hooghly was severely polluted, but denied that septic tank effluents had much to do with it. Pollution, according to government, was due to a combination of factors, including municipal sullage, boatmen and coolies working along the river, while septic tank effluents caused no or hardly any pollution. Thus, it denied the resolution without any detours. Nevertheless, the Council performed a fake vote in order to gauge opinion on the matter. The result of this exercise is noteworthy. Out of the 16 members who voted *for* the resolution, all were Indian (both Hindu and Muslim); out of the 18 members who voted *against* it, 15 were British. Clearly, river pollution policy in Calcutta continued to be significantly influenced by the close ties between British colonial officials and the European industrial lobby, and that lobby's own representatives in the colonial administration.[2]

Unable to ignore the issue completely, the Bengal government in 1919 agreed to undertake a full inquiry into the current state and sources of river pollution, and into the question whether legislation was required.[3] Over the next two years the director of public health, a Mr Bentley, conducted a full-fledged inquiry. However, by 1925 the Bengal Legislative Council still possessed no final report. Apparently, Bentley had gone to England, carrying with him all the material produced during the inquiry in order to finish his report while on leave. Increasingly impatient, Indian council members decided to form another committee of inquiry, but without any data at hand and the Bengal government unwilling to do anything until Bentley's return, the venture naturally failed. As a look at the Council proceedings suggests, Bentley's report was still pending as late as 1928(!). Considering the report's potential to instigate another fierce septic tank controversy similar to the one in 1904/05, the idea suggests itself that this delay was not merely coincidental, but another attempt by the Bengal government to conceal the truth about septic tank effluents. Occasional reports about bacteriological water and effluent analyses received by Council members during the early 1920s reveal that a disconcerting amount of effluent samples were grossly contaminated, ranging from 33 per cent to 75 per cent of all samples according to different reports.[4] The Bengal government, it seems, was well aware of the ongoing pollution caused by the mills' septic tank effluents, but still refused to take concrete steps to enforce the proper sterilisation of effluents as demanded by the Bengal Factory Rules.

In the United Provinces, the superintending engineer in 1926 drew a dismal picture of the existing municipal sewerage and drainage systems in the province:

> The maintenance of drainage works generally is defective [...]. In many places much money has been spent on the construction of sewers, the existence of which has even been forgotten. [...] No intelligent use is made of the existing drainage works, nor are they even maintained with care. [...] Where sewage farms exist the municipalities concerned do not take sufficient interest in the proper management of the farms; they are prone to

permit the cultivators to use the sewage in a reckless manner or to misuse it [...]. In some farms the rent realized is much lower that the real value of the land and the bulk of the sewage is allowed to run to waste.[5]

A major reason for this state of things, according to the director, was the absence of trained staff:

Local authorities generally are not able to obtain the services of competent technical officers, the reason being that the condition of service and method of work are not such as to attract men of standing. [...] The staff employed by municipal boards does not possess either the technical qualifications or the experience in most cases that the duties undertaken require.[6]

Evidently, the salaries municipalities were able to offer and the work conditions in general were unappealing to suitable candidates. But competent staff was not only missing at the municipal level, it was also scarce in the provincial public works department. The last recruitment of a new qualified sanitary engineer went back to 1920, and the present engineers complained that they were overloaded with work.[7] Complaints of this kind were nothing new. Already in 1908, the United Provinces sanitary conference noted that one sanitary engineer was in charge of the entire province, i.e. 89 municipalities, and had to fulfil not only advisory but also executive functions – 'an impossible task'.[8]

As colonial fiscal conservatism kept municipalities short of funds and trained staff, existing sewerage systems and sewage treatment facilities deteriorated rapidly, and river pollution continued and increased unabated.

In Banaras, the municipality's original plans to construct a sewage farm were never revived, and the city's untreated sewage kept being discharged into the Ganges above its confluence with the Varana. By the 1920s, this sewer and many of the smaller sewers had fallen into partial decay. As a result, sewage again ran into the Ganges over the bathing *ghāts*, much as it had before the original sewerage system was built in the 1890s.[9] In 1926, a group of Banarasi citizens formed the Kashi Tirth Sudhar Trust.[10] In its address to the Governor-General of India, the Trust drew attention to the urgent necessity of two public works: First, the repair of the *ghāts*, the foundations of which were being undermined by the scouring action of the river currents and were therefore in danger of falling in. Second, the construction of a sewerage system that would prevent the fouling of the Ganges 'by the indiscriminate discharge of sewerage', which at certain places had rendered the water close to undrinkable. Hence, the Trust requested the provision of competent engineers to investigate ways to train the river, repair the *ghāts*, and avoid the discharge of sewage along the riverfront. At the same time, it offered to collect the money needed to pay for this preliminary survey.[11]

The pollution of the Ganges along the city *ghāts* had engaged officials and the public for several years before the formation of the trust, but due to lack of funds nothing had been done. In reaction to the Trust's initiative, government officials at every level expressed their desire to see the sewers repaired. Yet, again, plans failed in the face of colonial fiscal policy. In 1928, the provincial government recommended that the sewerage project be financed by doubling the pilgrim tax, the returns of which would allow the municipality to raise a loan of 10 lakhs, and underlined that for many years to come no other source of funds would be forthcoming.[12] In reply, the Banaras municipal board noted that it was still facing a debt of 22 lakhs, borrowed for the original water and sewerage works.[13] Nevertheless, it agreed to the scheme and in December 1930 sent its resolution to double the pilgrim tax to the Government of India for sanction. Meanwhile, government-appointed engineers completed the preliminary survey at a cost of 50,000 rupees, fully covered by the Kashi Tirth Sudhar Trust.[14] The municipal board had made it clear that due to financial stringency, it was presently not in a position to contribute towards the survey.[15]

In 1935, more than four years after the board's request for sanction, the Government of India's decision regarding the doubling of the pilgrim tax was still pending. Writing to the commissioner of the Banaras division, one government official explained that the decision was being delayed due to opposition from the railway authorities. Moreover, other municipalities had sent similar requests, also still pending, which government viewed as more important. Therefore, the official cautioned, there was little hope for the Banaras municipality to have their own request granted.[16] As it appears, this turned out to be true. Neither the municipal files nor the annual reports of the director of public health consulted for this book contain any signs that the sewerage scheme was further pursued. The Kashi Tirth Sudhar Trust shifted its focus towards the preservation of the Tulsi Ghat in the years that followed.[17]

In Kanpur, the sewerage system was operative from the early 1900s onwards, with the intercepting sewer discharging sullage (including, amongst others, all the wastewater from the city's tanneries and other factories) untreated into the Ganges near the cantonment. For almost two decades, the location of the sewer outfall provoked indignant protests from the cantonment authorities, who considered it to be a menace to military health, and complained of the considerable financial losses it caused to the landlords who owned bungalows nearby.[18] The 'extremely offensive smell', one military officer wrote, was carried into the cantonment by the easterly winds and rendered several bungalows uninhabitable.[19] From 1914, the option of diverting the outfall to a sullage farm was discussed in earnest as part of the city's new sewerage and drainage scheme. The majority in the Kanpur municipal board, including the British chairman, were little enthusiastic about this proposal. The military authorities' complaints, they maintained, were exaggerated, and a sullage farm would require an enormous amount of land. However, some Indian members of the board did press for the construction of a sullage farm, claiming that the sewer outfall created a nuisance to bathers at the Bengali and Sidh Nath *ghāts*.[20] The discussion around the establishment of a sullage farm also brought to light the long-standing dominant attitude towards sewage treatment within the government of the United Provinces. Writing on Kanpur, the sanitary engineer in 1914 noted:

No doubt a time will come when, as the population on the river banks below cantonments increases, it will be necessary to subject the sewage to a process of purification before admitting it into the river. But at present there is no town of any importance for many miles below Cawnpore, and I think that the purification of the sewage, highly diluted as it will be with canal-water, is a contingency for which it is unnecessary to make provision in the present estimate. When however it becomes necessary to purify the effluent, one of the new bacterial processes can be adopted at a comparatively small cost, instead of an expensive sewage farm and pumping machinery, costing about 5 lakhs, and involving an annual expenditure on maintenance and pumping of over half a lakh of rupees.[21]

Considering the government's very recent failure in experimenting with biological sewage treatment, the sanitary engineer's statement seems very optimistic.

Kanpur's sullage farm plans were soon abandoned as the municipal board and the provincial government failed to reach an agreement over funding. Asked to cover one fourth of the costs for the sewerage and drainage schemes, the municipal board declined, citing its present financial situation and pointing out that there was also no prospect of obtaining funds via private subscriptions due to the depression in the markets.[22] The project was consequently shelved.[23] In 1920, the board drew up another comprehensive scheme for the sanitary improvement of the city, including, amongst others, a drainage project, a sullage farm and an infectious diseases hospital. Seven years later, it applied to the provincial government for a suitable loan and grant, but financial support was only sanctioned for the drainage project.[24] As pointed out by the United Provinces secretary, financial support for the sullage farm was out of question for the time being.[25] Later correspondence by the provincial public health department shows that several options for sewage treatment, including a sullage farm and experiments with activated sludge treatment, were further discussed,[26] but there are no signs that these proposals were effectively followed up.

Lucknow and other cities already possessing sullage farms fared no better. Lucknow's two farms, constructed after the failure of biological sewage treatment in 1909, soon proved insufficient to handle all the city sullage, and in 1914 their substitution by a new, bigger farm was proposed. After initial procrastination, the Government of India agreed to make a special grant of one lakh rupees, and by 1915, everything seemed set.[27] In 1918, the Lucknow Town Improvement Trust overturned all existing plans and called for a 'complete and modern system of drainage' for the city, demanding the remodelling of current schemes.[28] Over the next minimum 12 years, the central, provincial and municipal governments kept deliberating over funds, while spending some amount of money on project preparation, but nothing concrete materialised. By 1925, the old sullage farms showed first signs of decay. In one case, leaks appeared in the suction galleries at the pumping station, a matter that required immediate repair since the motors were under danger of being flooded. Shortly after, the tank sewer at the pumping station completely collapsed.[29] In 1930, the director of public health noted that 'the western intercepting sewer was choked with silt and was broken in places, allowing sewage to flow direct [sic] into the river Gomti [, and that] the sewage in Eastern intercepting sewer [sic] constantly overflowed through the manholes'.[30] Government records consulted contain no indication that the situation was improved up to 1930.

Even where sullage farms were still functional, they were not operated adequately, as the passage from the director's annual report cited at the beginning of this chapter has already revealed. In Lucknow, tenants apparently either took the sullage without paying for it, or refused to take it, allowing it to flow into the Gomti instead.[31] In Allahabad, sullage kept overflowing into the Yamuna from the very inception of the farm, and continued to do so up to the late 1930s (and, arguably, beyond).[32] The Agra sullage farm, too, was ill-maintained and by the late 1920s handled only a fraction of the city's wastewater.[33] A major reason for this state of things, according to the director, was the lack of skilled staff, which the municipalities blamed on financial stringency.[34] The economic situation during the late 1920s and early 1930s rendered improvements further unlikely. In his annual report of 1933, the director of public health noted:

> Retrenchments by municipal boards during the previous years continued during 1933 also. No new schemes of sanitary improvements were undertaken; in fact there is marked evidence of the normal expenditure under conservancy plans and appliances having been reduced in many cases below the previous year's working standard.[35]

Thus, river pollution was to continue unabated in the foreseeable future.

The United Provinces' and Bengal's lenient approach towards river pollution was matched by the Government of India, which continued to show little consideration for the issue. Debates around a national river pollution law in 1890 did not bear fruit, and it appears that national legislation was not seriously discussed anymore until Independence. The Indian government's committee on sewage disposal, established in 1893 to issue general guidelines, did advise strongly against the discharge of untreated sewage into rivers, stipulating that some form of sewage treatment – preferably through sewage farms – was mandatory. However, since the committee lacked support in official circles, these provisions were never enforced, and the state governments were ultimately left to frame their own policies on river pollution and sewage disposal. It was the independent Government of India which, in 1974, undertook the first extensive legal effort to tackle the colonial legacy of river pollution. The Water (Prevention and Control of Pollution) Act passed in that year vests state boards with regulatory authority and empowers them to lay down effluent standards, to inspect sewage and trade effluents as well as purification plants, and to evolve the most suitable methods for effluent treatment under local conditions. A central board coordinates the state boards' activities, and executes the same functions for Union territories.[36]

The major reasons for the colonial state's neglect, both at the central and provincial levels, were, for one, ideological. Initially, the Indian government's enduring opposition towards germ theory and waterborne disease aetiologies caused many colonial officials to deny the potential dangers involved in the discharge of untreated sewage into rivers. Considering that sewerage systems themselves were a brainchild of the miasmatic era, this close alliance between miasmatic beliefs and complacency towards river pollution comes as no surprise. Edwin Chadwick, the 'father' of modern sewerage infrastructures in Britain, in his *Report* of 1842 clearly maintained that the removal of putrefying wastes from urban surroundings was paramount to the cleanness of rivers.[37] From the perspective of miasmatic disease theory, streams appeared as viable sewage carriers. Proponents of sewage disposal into Indian rivers also built on the fact that germ theory during the early 1890s was still an evolving science. Numerous questions about germs and their behaviour under different conditions – e.g. how sewage and river water affected their survival and multiplication, or what combination of germs, predisposition and environment was necessary to cause an outbreak of disease – were still controversial, and bacteriology, in India as much as in Britain, was unable to provide clear guiding principles for policy makers involved with water quality issues.[38] The most crucial ideological construct shaping colonial river pollution policy, however, was what I have called the 'Indian paradigm'. As discussed in Chapter 2, the 'Indian paradigm' built on a long-standing colonial ideology that classified the Indian environment as 'tropical', and thus as inherently different from 'moderate' European environments,[39] an ideology which in turn can be situated within the overarching colonial ideology of difference.[40] According to the 'Indian paradigm', Indian rivers possessed self-purifying powers much greater than those of British rivers, due to their physical character, especially their water volume, and the alleged 'tropicality' of the Indian environment as a whole, including the hot climate and the strength of the 'tropical' sun. Sewage disposal into Indian rivers, it was therefore believed, was far less problematic than sewage disposal into British rivers. Applied within the context of the miasmatic disease theory, the 'Indian paradigm' promised the rapid destruction of putrefying organic matter once it entered a river. Applied within the new context of germ theory, the same paradigm allowed colonial officials to claim that Indian rivers possessed enhanced germicidal powers. Thus, the 'Indian paradigm' spanned the epistemological shift occurring in official circles with regard to disease theory during the 1890s, and even though it was in no way uncontested, it had turned into the major guiding principle for policies on sewage disposal into rivers by the end of the nineteenth century. As depicted in Chapter 5, the paradigm's influence did not remain restricted to the North-Western Provinces alone, but was similarly formative for policies in Bengal. With reference to the 'Indian paradigm', sewage and septic tank effluent disposal into rivers were generally viewed as unproblematic, except where a river's water volume was very low.

A key figure in the consolidation of the 'Indian paradigm' during the latter part of the 1890s was Ernest Hanbury Hankin. Hankin, director of the North-Western Provinces' bacteriological laboratory in Agra since 1892, was the first expert bacteriologist to conduct research into the self-purifying properties of Indian rivers.

His experiments on the Ganges and the Yamuna, which according to him not only proved the Indian climate's conducive influence on river self-purification, but also the existence of a mysterious germicidal agent in these two rivers, lent scientific support to the paradigm and thus confirmed the North-Western Provinces in their approach. Hankin's work was not only highly influential for the further course of government policy, it also presents a watershed in the development of colonial expertise on river pollution and sewage disposal. Hitherto, this field has not been addressed by the historiographies on science and technology in British India. While the historiography on technology has largely ignored sewerage systems,[41] the historiography on science has equally overlooked related developments, laying emphasis, among others, on 'natural history sciences' such as botany, mineralogy and geology (pursued by colonial amateur scientists from an early date), as well as medical science.[42] Arguably, this neglect – similar to South Asian environmental history's neglect of urban environments – mirrors the colonial state's own priorities, which determined the amount of administrational records produced on a subject and thus to a certain extent direct historians' interests today. While the present scope does not allow an exhaustive analysis, certain insights derived from the previous chapters may be summarised here:

As discussed earlier, metropolitan expertise on sewage disposal into rivers and related questions of water quality played a fundamental role in informing official opinion in India during the early 1890s. At the time, the colonial state had no experience with the subject, and only a small number of highly trained bacteriologists were present among its own ranks. The recourse to metropolitan expertise was mostly indirect, namely through reference to publications of experts and commissions prominently involved in British river pollution debates, such as Edward Frankland, Charles Meymott Tidy and the Rivers Pollution Commission. Since metropolitan debates themselves were marked by strong disagreements over disease theory, river self-purification, water analysis and many other relevant issues, the Indian debate took on equally controversial forms, with both advocates and opponents of the Banaras sewerage project selectively drawing on a range of contradictory expert opinions to support their argument. Ultimately, the North-Western Provinces under the suasion of the 'Indian paradigm' decided that the question called for an empirical investigation on the spot. The only metropolitan expert consulted directly for this was the British sanitary engineer Baldwin Latham, who stayed in India in 1890/91 to advise several cities on their sewerage systems.[43] However, Latham's main task lay with the technical aspects of construction, and his opinion with regard to sewage disposal into the Ganges was not based on an actual scientific inquiry. At the same time, two investigations were carried out under the direction of the government's own sanitary and engineering staff. A first, arguably impressionistic, survey aimed to assess to what extent riparian populations in and around the provinces' major cities used river water for drinking, and whether they considered it to cause sickness generally or at certain times of the year. The second, more scientifically grounded, investigation was a series of chemical water analyses taken of river water at Kanpur and Lucknow, executed by the resident engineer of the

Lucknow waterworks. According to critics, both these ventures were essentially flawed, since they did not include bacteriological water analyses, and since the present discharge of wastewater could not be compared with the organised discharge of sewage as envisioned by the Banaras and other sewerage projects. From a historical perspective, the North-Western Provinces' early inquiries are symptomatic for the enduring backward state of scientific research in the colony until the late nineteenth century. India had remained isolated from modern science for much of the century, and it was only under Lord Elgin's governorship (1894–99) that the colonial state started to make serious efforts to catch up.[44] Elgin's complaint that scientific investigations in British India were conducted by officials 'who are neither by experience nor knowledge competent to offer a decided opinion as to the best course to be pursued'[45] seems rightly placed when considering the river pollution debates and the North-Western Provinces' two related inquiries of the early 1890s.

The establishment of a bacteriological laboratory in Agra in 1892, and the appointment of E.H. Hankin, who had previously worked under Koch and Pasteur, as its director, came as part of the growing official acceptance of bacteriology in British India in the course of the 1890s.[46] With Hankin, river pollution research for the first time was put on a solid scientific footing, as both the chemical and bacteriological aspects of water quality were incorporated. Hankin himself, it seems, did not continue his studies after being appointed as a member of the Indian plague commission in 1896.[47] However, with the introduction of biological sewage treatment in India at the turn of the twentieth century, a number of colonial officials started to investigate different methods of sewage disposal and related questions of river pollution, sometimes in close interaction with metropolitan experts. Hankin's research was most directly continued by the Septic Tank Committee in Calcutta, which investigated the self-purifying capacity of the Hooghly in relation to septic tank effluents, as well as different methods of effluent treatment. A direct follow-up to the Committee's work was the joint inquiry by Bengal's sanitary commissioner William Wesley Clemesha and British sewage treatment expert Gilbert J. Fowler. This was continued by Clemesha for many years after Fowler's departure, and led to the establishment of a special laboratory in Calcutta for the purpose. In 1910, Clemesha published one of the first manuals on sewage disposal in the 'tropics', dealing with different methods of biological sewage treatment, in particular septic tanks, and addressing the vexed question of sewage disposal. As to the discharge of sewage and septic tank effluents into Indian rivers, Clemesha took on a rather cautious approach, considering the practice as a great potential danger.[48] A much more exhaustive treatise on sewerage and sewage treatment in the 'East', with some reference to sewage disposal into Indian rivers was produced some years earlier by Charles Carkeet James, sanitary engineer to the Bombay municipality.[49] These two were followed by a third major contribution on the subject in 1924, George Bransby Williams' *Sewage Disposal in India and the East*.[50]

The work of Carkeet James, Clemesha, Williams and other colonial officials raises a range of questions. For example, what kind of networks – within British

India, between colony and metropole, within the British Empire and beyond –
emerged that advanced and shaped colonial expertise on the construction of sewerage systems, various methods of sewage disposal, and sewage treatment in
British India? The importance of transnational networks for the generation and
circulation of knowledge has been extensively explored, for instance, by scholars
writing on nineteenth-century botany and other natural sciences, and needs to be
investigated for this topic, too.[51] At the turn of the twentieth century, colonial
officials in various Indian provinces experimented with different methods of biological sewage treatment, while in some places, such as Madras, crop irrigation
with sullage was already practised for many years.[52] At least some of these officials were actively involved with sewerage projects in other 'Eastern' countries
as well. Thus, the Egyptian government in 1902 invited C. Carkeet James to
Alexandria to receive his opinion on the sewerage system devised for that city.[53]
Metropolitan experts such as Baldwin Latham and Gilbert J. Fowler visited a
great number of Indian cities for consultation. In 1916, Gilbert J. Fowler, who
was originally based in Britain and had worked at the Lawrence experimental
station in Massachusetts for some time, was appointed professor of applied
chemistry at the Indian Institute of Science in Bangalore. At the Institute, he
continued his research on the activated sludge process of sewage treatment, and
later took the post as principal to the Harcourt Butler Technological Institute in
Kanpur. Fowler remained in India until his death in Bangalore in 1953.[54] Carkeet
James and Fowler may offer particularly fruitful case studies for future scholarly
investigations into the development of 'sewage expertise' in British India, the
transnational networks within which it evolved, and the reception of colonial
contributions by professional communities outside British India. In relation to
the last point, Fowler's presentation on septic tanks in front of the International
Congress of Hygiene and Demography in Berlin in 1907, and the reception of
the Septic Tank Committee's research in Calcutta by the scientific community
outside British India (referred to in Joseph Race's publication on chlorination),
may provide useful starting points.[55]

Another event that calls for closer attention in this context is an exchange
between the British Royal Society's Indian Advisory Committee (IAC) and the
Indian Board of Scientific Advice (BSA), the two main bodies created at the turn
of the twentieth century to coordinate scientific research in India.[56] Around 1904,
the two arranged for an interchange of views and reports between men engaged
with research on different methods of sewage disposal and treatment in Britain
and India.[57] In response, the Government of India sent a collection of reports
produced by different provincial governments to Britain.[58] The sources consulted
for this book do not suggest that any enduring formal cooperation, either
between different Indian provinces or between Britain and India ensued from
this initial exchange. As Deepak Kumar has noted, scientific research in British
India during the early twentieth century overall remained contained within the
provinces, and the latters' primary interest lay with immediate practical results
rather than research.[59] However, other sources, e.g. files of India's revenue and
agriculture department, may contain information to the contrary. Also, the files

of the British Royal Commission on Sewage Disposal may furnish insights into whether and how Indian reports were utilised by that commission during its investigations.

The second major factor responsible for the colonial state's neglectful attitude towards river pollution, next to the 'Indian paradigm', was fiscal conservatism. Generally, the colonial state's fiscal policy was oriented along the maintenance of its military and civil institutions, whilst urban development was given little importance. On these grounds, the central and provincial governments in the 1870s devolved the financial responsibility for urban development to the municipalities, without however creating a structure that would have allowed the latter to raise adequate municipal incomes. Dependent on revenues mostly derived from the octroi tax, and relying on the central and provincial governments' willingness to sanction loans and grants-in-aid, municipalities remained dramatically underfunded, and therefore ill-equipped to pursue expensive infrastructure projects. Funding opportunities such as public borrowing, which enabled contemporary European and US cities to realise far-reaching programmes of urban reform, were unavailable in British India.[60] As a case in point, Paris invested over 40 million pounds into urban development during the late nineteenth century, topped by an additional 35 million for continuing measures. In the US, the central government approved a budget of over 10 million pounds for one single sanitation project in Washington. Calcutta, in contrast, spent a mere 6 million pounds during the period. Thus, colonial fiscal policy laid the cornerstone for the chronic underdevelopment of Indian cities, which is still palpable today.[61]

In the United Provinces, fiscal conservatism directed official decisions about sewerage technology at every step. During the 1890s, the Banaras and Kanpur municipal boards were forced to drop their initial plans to construct sewage farms as part of their sewerage systems due to lack of funds. Unable to raise municipal taxes further in view of public protest, and not receiving any meaningful concessions from the central and provincial authorities, Banaras by the turn of the twentieth century stood at the verge of bankruptcy, while the Kanpur municipal board pulled the emergency brake by withdrawing its sewerage project as a whole. Consequently, untreated sewage (Banaras) and sullage containing great amounts of industrial effluent (Kanpur) kept being discharged into the Ganges unhindered. In the colonial discourse, fiscal conservatism was of course closely interlinked with the ideology of the 'Indian paradigm': The construction of expensive sewage treatment facilities, many argued, was unnecessary since Indian rivers rapidly purified themselves anyway.

Just a couple of years later, fiscal conservatism contributed significantly to the failure of the experimental sewage treatment station at Lucknow, as the Secretary of State for India, and subsequently the central and provincial governments, refused to sanction funds for the permanent employment of a 'sewage specialist'. This decision had far-reaching implications for the development of sewage treatment and river pollution in the United Provinces. Not only did it undermine the continuation of the provinces' research into biological sewage treatment, it also left them without sufficient numbers of skilled staff to oversee municipal sewerage systems and sewage/sullage farms where they existed. Twenty years later, the United Provinces' leading public health officers and engineers blamed the decay and ill-maintenance of municipal sewerage infrastructures primarily on the lack of skilled supervision. Financial stringency on a more general level also precluded the necessary financial investments for the material maintenance of existing sewerage infrastructures.

In Calcutta, the Bengal government's septic tank effluent policy was less shaped by colonial fiscal conservatism. This, however, is self-explanatory, since the financial responsibility for the mills' sanitary arrangements lay with the mills themselves. In a way, the mill managers displayed their own 'fiscal conservatism', as they vehemently defended the use of septic tank latrines and the discharge of septic tank effluents into the Hooghly simply because it was the cheapest and most convenient arrangement for them. The Bengal government's strong support of this practice fed on several motives. For one, the 'Indian paradigm' held considerable influence in official circles. Second, septic tanks were woven into the colonial discourse of 'modernisation' and 'progress', a discourse that had first emerged under the utilitarian government of Governor-General William Bentinck (1828–35) and was moulded specifically around the introduction of Western technologies in India under Governor-General Lord Dalhousie (1848–56). Under this ideology, the British saw it as their mission to 'modernise' and 'improve' the allegedly backward Indian society, an argument which was aptly used as a justification for colonial rule.[62] In the urban context, the replacement of traditional sources of water supply with a centralised piped water supply, and the introduction of sewerage systems in place of dry conservancy was viewed as an integral part of urban modernisation.[63] Thus, Bengal officials repeatedly stressed that septic tanks were more sanitary and 'modern' than hand removal and trenching, and therefore to be strongly promoted by the government. However, beneath this rhetoric lay more practical interests as well, since many officials viewed waterborne sewerage systems and biological sewage treatment as an opportunity to get rid of government's long-standing, precarious dependency on Indian sweepers.[64]

Third, and most importantly, Bengal's septic tank effluent policy was decisively shaped by the presence of two major local pressure groups: The predominantly European industrial lobby, and the Western-educated, predominantly Hindu, Bengali elite. Out of these two, the European industrial lobby exerted the most direct influence over official policy. Even though the Bengal government did make successive efforts towards the regulation and supervision of septic tank installations and effluent sterilisation, the official record clearly suggests that it retained a fundamentally pro-industrial mode and that the enforcement of provisions was weak. This way of handling the septic tank effluent issue, I argue, was part of a dynamic that developed between the Bengal government and Calcutta's industrial lobby over questions of industrial pollution during the late nineteenth and early twentieth centuries. Like so many aspects of urban environmental history, this topic has remained nearly unexplored for British Indian cities[65] and calls for many much more detailed case studies.[66] The following paragraphs may present a starting point for this.

In late nineteenth and early twentieth century Calcutta, at least two other debates occurred over industrial pollution. For one, Indian members of the Calcutta Corporation during the early 1890s raised objections against the discharge of factory effluents into the Hooghly by the Titagarh and Kankinara paper mills, which they feared threatened the city's water supply.[67] The Bengal government's initial response was very supportive. With the aim to control river pollution from an early stage, the Lieutenant-Governor of Bengal proposed a bill that would prohibit the discharge of 'any solid or liquid sewage matter, or any poisonous, noxious or polluting matter, whether liquid or solid, proceeding from any factory or manufacturing process'.[68] However, the bill was withdrawn after consultation with the Bengal Chamber of Commerce, and an arguably biased inquiry into effluent discharge by the government's special inspector of factories, which relied simply on visual observation and the mill managers' own views. Considering the close ties between the Bengal government and Calcutta's European industrialists at the time,[69] the government's abrupt change of mind was quite obviously brought about by the industrial lobby's successful intervention.

A second case of industrial pollution, engaging the Bengal government from the mid-nineteenth century until the end of colonial rule, was air pollution.[70] The opening of India's first railway line between Calcutta and the Raniganj coal fields in 1855 propelled a massive increase in the number of coal-fuelled steam engines in the city, employed by ships, railway locomotives, engineering workshops, the government mint, municipal water pumps, and the growing number of jute mills and other factories. As a result, numberless chimneys shrouded Calcutta with dense black smoke, which soon transformed into recurring and worsening smogs due the city's geographical setting and meteorological conditions. Under the persistent pressure from European and Bengali elites (professionals, administrators and mercantile groups in particular), the Bengal government took on an increasingly firm stance against what contemporaries referred to as 'smoke nuisance'. The Calcutta and Howrah Smoke-Nuisances Act of 1863 – one of the earliest air pollution laws in the world, modelled after the one London had passed ten years earlier – demanded that the 'best practicable means'[71] be used to prevent smoke emissions. However, the Act overall remained a dead letter. Under considerable public pressure induced by a serious bout of smog in Winter 1878/79, the Bengal government in 1880 called for a special committee, and following the latter's recommendations, took steps towards a more rigorous implementation of the Act. To this end, it appointed a special inspector with the task to supervise boilers and inspect smoke emissions from chimneys, to issue warnings to mill managers, and initiate prosecutions.[72] At the same time however, government retained a fundamentally pro-industrial approach, relying on consultation and advice rather than immediate prosecutions, especially when the mill in question was European-owned. Securing a conviction, moreover, was difficult. Mill managers resisted smoke abatement strongly, citing financial and technical reasons. From the 1890s, they successfully fought against prosecutions in court, and within a few years, the magistracy came to firmly side with them. A major problem was the Act's use of the wording 'best practicable means', which left ample room for interpretation and made it difficult to prove any concrete neglect by the mills. Well aware of this, many mill managers felt safe enough to ignore the smoke inspector's notices.

Calcutta's anti-smoke lobby finally got the upper hand following the appointment of the autocratic Lord Curzon as Governor-General of India in 1899. Curzon took strong interest in the matter, and on his initiative, the smoke inspector for Leeds, Frederick Grover, was invited to Calcutta in 1902 to report on the 'smoke nuisance'. Grover's inquiry directly led to the Bengal Smoke Nuisances Act of 1905, which created the necessary legal framework to intensify monitoring and regulation. Amongst others, the Act provided for the establishment of the Bengal Smoke Nuisance Commission, a body that was to supervise the work of a permanent smoke inspectorate led by a chief inspector. Moreover, violations now were defined along actual smoke emissions – measured on a daily basis by use of a standardised procedure to assess the density and duration of smoke – rather than along firing practices and technologies. At the same time, great efforts were made to secure the cooperation of the reluctant industrialists. Most importantly, the Bengal Smoke Nuisance Commission was constituted of an equal number of government officials and industry representatives, drawn from the Bengal Chamber of Commerce and the Bengal National Chamber of Commerce. The Bengal Chamber of Commerce was moreover consulted during the drafting stages of the Smoke Nuisances Act, and was able to obtain important concessions, for example a massive reduction of the first conviction fine from 1,000 to merely 50 rupees.

Before long, the Smoke Nuisance Commission's work came to be marked by the heavy representation of industrial interests in it. After a short period that saw a number of prosecutions, the Commission in 1912 adopted a new system that factually transferred the accountability for smoke abatement from the mill managers to the stokers. On observing excessive smoke emissions, the smoke inspector now sent notice to the mill managers, who then levied a fine on the allegedly negligent or inefficient stoker. This policy was not limited to factories, but came to be applied equally on ships, locomotives and elsewhere. Thus, the industrialists and the government achieved a compromise which 'shifted the moral and physical burdens of smoke abatement onto labour',[73] creating a highly coercive system that ensured smoke abatement while simultaneously taking responsibility off the mill managers. This policy proved highly efficient. According to the smoke inspector, the chimneys responsible for the worst emissions in 1906 cut the latter down to a mere 11 per cent by 1916. In 1922, the Smoke Nuisance Commission was in a position to declare that Calcutta's air was cleaner than that of European industrial and shipping centres, even though the number of mills had doubled since 1907.

Overall, the Bengal government pursued industrial air pollution regulation much more keenly than industrial river pollution regulation (be it through industrial wastes or septic tank effluents) during the early twentieth century. By 1926,

Calcutta operated the most ambitious, comprehensive and efficient smoke pollution control system in the world.[74] By that same year, reports on the water quality of the Hooghly were as alarming as they had been in the 1900s, Indian members of the Bengal Legislative Council were still waiting for the results of a river pollution inquiry conducted in 1919–21, the publication of which was arguably retarded on purpose, and there are no signs that the Bengal government was willing to undertake any serious effort to enforce septic tank effluent regulations in the foreseeable future. What are the reasons for this discrepancy?

Generally, all the three individual cases under discussion here betray one major commonality, namely the Bengal government's fundamentally pro-industrialist approach. Consultation and advice, rather than prosecution, were the government's major guiding principles, both while framing environmental regulations on industrial pollution and while implementing them. The withdrawal of the government's initial proposal for a river pollution law in 1895 is a case in point. Another case in point is the fact that government never forced the mills to use septic tank effluents as boiler feed-water. This was successfully practised by factories in Britain at the time, and was recognised as a very feasible alternative by the Septic Tank Committee, the Lieutenant-Governor of Bengal and other officials from 1904 onwards. Instead, government accepted the mill managers' preference of having a special inspector of septic tank installations appointed.[75] A well-founded evaluation of the inspectorate's work can only be given after more thorough research into individual cases. However, the sources consulted for this book bear concrete proof that at least during the years around 1910, the majority of mill managers were not acting upon the multiple warnings issued by the septic tank inspector, and that the Bengal government was reluctant to initiate actual prosecutions. In 1910, Bengal's sanitary commissioner reiterated his earlier complaints that most septic tank installations were not given adequate attention by the mill managers, showing various deficiencies in operation and maintenance, and that the chlorination of effluents was done satisfactorily only by very few. Consequently, he urged that 'the time has arrived when disregard of regulations should meet with suitable punishments'.[76] In response, the Bengal government maintained that '[p]rosecution need only be resorted to in the event of continued failure to bring the installation up to a proper standard'.[77]

The major reason for the discrepancy between air pollution and river pollution regulation enforcement, then, appears to be the specific compromise forged by government and industrialists in 1912, which guaranteed smoke abatement without putting actual responsibility on the mill managers. Quite obviously, the Bengal government was not willing or able to take a firm stance against the managers themselves, no matter what form of industrial pollution was concerned. By transferring the responsibility for smoke abatement onto the stokers, government was able to satisfy its own and the mill managers' interests at the same time. No similar compromise was pursued that would have guaranteed the enforcement of septic tank effluent regulations. Generally, it seems that air pollution was of much stronger concern to Calcutta's European elite than river pollution. Thus, movements against air pollution involved Europeans to a significant extent and received considerable official backing time and again (e.g. through Lord Curzon),[78] while public resistance against septic tank effluents was mounted entirely by Indians. In this context, it needs to be remembered that air pollution affected *all* the citizens of the capital, whereas polluted river water mostly affected those who had no access to filtered municipal water supplies – i.e. the poorer Indian classes – and those who resorted to the river for ritual and bathing purposes – i.e. Indians, especially Hindus.[79] Finally, the 'Indian paradigm' exerted considerable influence among colonial administrators, which generally attenuated fears over river pollution.

The cooperation between the Bengal government and Calcutta's European industrial lobby is a perfect example for how colonial policies on river pollution were actively shaped by the political, social and economic contexts prevailing at the 'ground-level' of the city. The detailed case studies of Calcutta and Banaras discussed in this book have revealed the presence of another major pressure group: Indian, and more particularly, Hindu citizens. Scholars writing on post-Independence debates around river pollution in India maintain that Hindu religious perspectives on the Ganges and related religious notions of purity and pollution have undermined the success of environmental campaigns. Most Hindus, they contend, translate the Ganges's spiritual power of purification to the material plane, holding that the goddess is capable of removing material pollution while her own sacred purity remains untouched by it.[80] In contrast, this book shows that religious perspectives around the turn of the twentieth century tended to initiate and reinforce movements for keeping the Ganges free from sewage, rather than weaken them. A considerable number of Hindu citizens in Calcutta and Banaras – including administrators, riparian residents and the press – challenged official river pollution policies, as they considered the discharge of sewage and septic tank effluents into the Ganges (generally, or at certain locations) not only as a potential health hazard, but also a religious sacrilege due to the great amounts of excreta contained in them. That said, it needs to be stressed that Hindu opinion on the matter was not homogenous. For certain Hindu members of the Banaras municipal board, for instance, wastewater removal from the city took priority over religious considerations.[81] In Calcutta, some of the Hindus most actively fighting against septic tank effluent disposal into the Hooghly did so entirely on public health grounds.[82] To be born Hindu (or into any other religion, for that matter), then as now, does not mean that people actually hold on to respective beliefs, and interpretations of the latter may vary significantly among different persons.

In Calcutta, Indian resistance was particularly vigorous due to a combination of political, social and environmental factors that rendered situations in the capital unique. For one, the city's Western-educated, predominantly Hindu Indian elite had developed a particularly keen political awareness over the nineteenth century, with attitudes growing increasingly critical towards the colonial state from the mid-nineteenth century onwards. By the early 1900s, Bengal was a hotbed for different strands of Indian nationalism and revolutionary movements, which meant that the colonial state's actions stood under sharper scrutiny than ever before.[83] Second, Indian resistance in Calcutta was most certainly intensified by the fact that the mills discharged their septic tank effluents above and within city limits, starting with mills located as far upstream as Kankinara. This directly affected the thousands of Indians bathing in the river daily and drinking water unfiltered from it, and also put into question the safety of filtered municipal water supplies. Moreover, the Hooghly at Calcutta is tidal, which means that any waste put into it is bound to return with the next flood tide. In contrast, Banaras and Kanpur discharged their untreated sewage and sullage *below* city limits, thereby actually contributing to the immediate cleanliness of the bathing *ghāṭs* and the Ganges running by the city. Significantly, the Kashi Ganga Prasadini Sabha's initial *raison d'être* was not the prevention of sewage disposal into the Ganges per se, but the cleanness of the 'Ganga of Kashi', i.e. the stretch flowing within the sacred limits of the city marked by the rivulet Assi to the south and the stream Varana to the north. Thus, the exact points of sewage discharge and local environmental settings played an important role.

In both cities, Indian lobbying and protest were initiated and led by local Hindu elites. In Calcutta, protest against the Bengal government's septic tank policy united liberals and conservatives, as evidenced by the wide spectrum of Indian newspapers supporting the agitation.[84] While the dynamic between Calcutta's leading class and other segments of society needs to be explored further in order to achieve a concise picture, the Rishra petition and reports on a protest meeting of 500 people at Konnagar do suggest that the septic tank controversy mobilised a wider populace, including non-elite Hindus.[85] In Banaras, the Kashi Ganga Prasadini Sabha was made up entirely of members of the city's elite, many of whom served as municipal commissioners as well. The same is true for the Kashi Tirth Sudhar Trust, who followed the Sabha's footsteps in 1926. The Sudhar Trust differed from the Sabha insofar as it was not an exclusively Hindu body, counting several Muslims among its members.[86] There are no signs that popular protests and riots against the Banaras waterworks and sewerage projects were motivated by the proposed discharge of sewage into the Ganges. Instead, popular discontent was primarily fuelled by the financial pressures these projects put on the citizens by way of increased taxation during a time of general economic hardship. Nevertheless, official records do contain some reference to strong 'public sentiment' against the practice,[87] and it is quite possible that this may have involved non-elite citizens as well.

The analysis of Calcutta's and Banaras's 'environmental pressure groups' supports David Arnold's recent contention that the social profile of urban environmental activism in British India differed significantly from that of its rural counterpart. In rural areas, colonial environmental policy (especially forest legislation) stirred the poorer classes into a struggle for the preservation of their livelihoods. In contrast, urban activism against air and water pollution 'was most marked among the higher castes and middle classes seeking to protect their neighbourhoods or to defend their religious sensibilities'. Writing on Calcutta and Bombay, Arnold finds little evidence for complaints issued by the urban poor.[88] Rural environmental movements of the nineteenth and early twentieth centuries therefore can be viewed as early precursors of today's 'environmentalism of the poor',[89] while urban environmental pressure groups anticipate contemporary Indian 'bourgeois environmentalism'.[90]

Indian pressure groups in Calcutta and Banaras, despite their sustained lobbying and agitation, were not able to exert any lasting influence over official river pollution policy. At best, they were able to achieve some temporary concessions, simply because the colonial state could not ignore them completely. The North-Western Provinces in acknowledgement of 'public sentiment' came round to support Banaras's sewage farm project for a while, a support that was against the personal inclination of Sir Auckland Colvin and other British officials. However, religious arguments ultimately carried little weight, and failed to alter finance-driven colonial agendas. In Calcutta, the Septic Tank Committee explored the motives of Indian, and more particularly Hindu, protest by consulting Hindu lay 'witnesses' and religious scholars, but its proceedings clearly reveal that the inquiry's major aim was to use their statements to invalidate religious opposition as a whole. In its final report, the Committee maintained that Indian protest was based on mere 'sentiment', and therefore was not to have any impact on future government policy. During the decades that followed the Committee's inquiry, Indian pressure groups came to exert a strong *indirect* influence on government policy. The potential threat of renewed agitation caused the Bengal government to adopt a policy of censorship, in which the results of river pollution inquiries were actively covered up. Thus, Bengal intentionally delayed the publication of Gilbert J. Fowler's unfavourable septic tank report of 1906 for many years, and caused Fowler to omit any direct reference to his work in Calcutta from his presentation given in front of the International Congress of Hygiene and Demography one year later. Similarly, government at least until 1928 held back the results of the Hooghly river pollution inquiry of 1919–21.

A closer look at the Calcutta and Banaras river pollution controversies lays open a deep ideological rift between British colonial officials and local Indian pressure groups. Indian opposition to varying degrees built on a Hindu perspective on the river and Hindu concepts of ritual 'purity' and 'pollution', while British officials held an entirely secular view on the river and defined 'purity' and 'pollution' along contemporary standards of bacteriology and chemistry.[91] With their argument, the Hindus not only fundamentally challenged government policy as such, but also the sole validity of the secular worldview on which it built. In their view, religious perspectives carried equal, if not more, importance. That said, it needs to be stressed that Indian opposition, especially in Calcutta, was not exclusively based on religion, but was driven by strong public health concerns (and maybe underlying political motives) as well. Western science was well established in early twentieth-century Bengal and Indians were quite aware of the implications a bacteriologically or otherwise polluted Hooghly may have for public health. Significantly, they utilised the septic tank controversy as an opportunity to vent their long-standing frustrations over other forms of industrial river pollution.

The Septic Tank Committee's interrogation of Hindu 'witnesses' presents a particularly vivid example of how the colonial state and its Hindu subjects negotiated over these different meanings of 'purity' and 'pollution'. In particular, it shows how the state tried to lay down certain definitions of 'purity' and 'pollution' as authoritative that would serve its own practical interests. In this, the Committee's proceedings bear strong similarities to colonial discourses around well water pollution during the nineteenth and twentieth centuries. In her analysis of court files produced around well water disputes, Kelly D. Alley demonstrates that there existed a strong discrepancy between Hindu, caste-based notions of water pollution (i.e. the ritual pollution of well water committed by the touch of lower castes), which was held by landowners and upper-caste Hindus controlling the access to these wells, and the secular notion of water pollution as held by British colonial officials. Alley argues that, instead of directly challenging the proprietary rights of upper castes to well water, the colonial state sought to undermine these rights by advancing a new definition of water purity and potability, investing terms like 'pollution', 'contamination', 'fouling' and 'defiling' with a thoroughly secular meaning.[92] Viewed in light of each other, the Septic Tank Committee's proceedings and the colonial judiciary's approach towards well water disputes point to a more general colonial strategy in which the state sought to undermine religious concepts of water pollution in order to fulfil its own interests and expand its hegemony. The Septic Tank Committee's approach was particularly ingenious in this regard. The Committee, instead of openly antagonising the Hindu community by denying the validity of Hindu notions of 'purity' and 'pollution' as such, chose to put forward a *specific* interpretation as authoritative – one that stressed the immutable purity of the Ganges – that would justify its septic tank effluent policy and invalidate Hindu religious opposition.

In sum, colonial river pollution policies during the late nineteenth and early twentieth centuries were framed entirely according to the colonisers' own cultural paradigms, and failed to incorporate Hindu religious views on the Ganges.

The colonial state's disregard for religious perspectives on the river and related notions of 'purity' and 'pollution' presents an interesting connection to the Government of India's river clean-up programmes after Independence. Until very recently, government river clean-up programmes such as the Ganga Action Plan (GAP) have failed to take into account Hindu religious perspectives on the river, which according to many scholars and environmentalists tend to obscure the problem of environmental pollution and hinder the development of public environmental awareness. Thus, the GAP built on a singularly scientific worldview which is not accepted or understood by the majority of the people.[93] A major difference between the colonial and post-colonial periods of course lies in *how* religious perspectives acted on popular environmental awareness. In colonial Banaras and Calcutta, religious perspectives on the river initiated and reinforced Indian movements against river pollution, whereas at least since the 1980s they appear to have fostered environmental complacency. What are the reasons for this difference? Are the Banaras and Calcutta controversies at all representative for other colonial river pollution debates in this regard? If yes, when and why did Hindu perspectives start to operate *against* public environmental awareness? These questions, tantalising as they are, can only be answered on the basis of further research.

The above-mentioned continuity is just one among many that links independent India's struggle with river pollution to colonial precedents. On a general level, India has perpetuated a range of structural deficits responsible for the underdevelopment of colonial cities. Among these, most importantly, stands the lack of municipal administrational and financial autonomy. While the colonial state granted municipal boards the power to raise taxes and found administrative institutions, the actual control of finances remained in the hands of the British.[94] Thus, the Banaras municipal board in the early 1930s was unable to double its pilgrim tax – aimed at generating income for the prevention of river pollution and general town improvement – because the Government of India refused to sanction the proposal.[95] This system of external political and financial control over Indian cities is still in place today. Governors appointed by the central government as constitutional heads for each federal state have the power to directly interfere into municipal affairs, particularly with regard to financial matters. The administrator in Delhi, who is viewed as the direct successor of the colonial commissioners, is authorised to overrule decisions of various committees on behalf of the central government. In addition, there existed and do exist a number of bodies and institutions with overlapping competencies, which further complicates urban planning. The Government of India's 74th Constitutional Amendment Act of 1993 vested municipalities with more administrational and financial autonomy for the first time after Independence, but in practice, little has changed since then. As a case in point, a mere 3.4 per cent of municipal budgets was drawn from municipal taxation at the beginning of the twenty-first century.[96]

Government river clean-up programmes have been similarly hampered by structural deficits. In 1985, the Government of India created the Central Ganga Authority (CGA).[97] The CGA's task was to supervise and guide the Ganga Project Directorate (GPD),[98] a special body within the Ministry of Environment and Forests that was to sanction and coordinate GAP activities. The Central Pollution Control Board (CPCB) acted as the major executive organ in this constellation, supervising the state pollution control boards, which administer pollution laws on the state level. More generally, the CPCB is responsible for the implementation of the Indian government's environmental objectives and planning, and for the enforcement of existing environmental legislation. The involvement of numerous bureaucratic institutions with often overlapping competencies led to frequent conflict, and this created serious stumbling blocks. Moreover, the fragmentation of responsibilities and accountabilities onto different agencies promoted widespread corruption and nepotism. For instance, the Haryana State Pollution Control Board was dissolved in 1992, shortly after it had initiated prosecution against the Haryana Chief Minister's son-in-law. In Bihar, the state government kept its Pollution Control Board seriously underfunded for several years, restricting expenditure to one third of the Board's requisition. As a whole, the GAP's administrational design contributed significantly to the executive's persistent failure in implementing its aims.[99] Faced with the government's dismal performance, activists started to take recourse to public interest litigation at the Indian Supreme Court and the High Courts. Through this, a new pattern of judge-driven implementation emerged in which the Courts took over the role of the Pollution Control Boards, keeping up sustained pressure on polluters, enforcing effluent treatment, and directing the closure of industries that failed to comply.[100] By the new millennium, river clean-up programmes had become 'inextricably tied to a maze of institutions, bureaucracies, and court orders'.[101]

India's environmental policy as a whole is still firmly linked to colonial precedents in many ways. Forest and water management policies, for instance, strongly reflect the colonial state's centralised and technocratic approach. Forest bureaucracies continue to be guided by the framework provided by the Forest Act of 1927, and the centralised policing regime set out by the Forest Act of 1980 continually undermines local participation in forest management.[102] India's first multi-purpose river valley development projects were introduced by the colonial state in the 1940s and have been aggressively advanced and implemented by independent governments since then (e.g. the Sardar Sarovar project in Central India, or the massive Tehri Dam in Uttarakhand).[103] Most recently, Prime Minister Narendra Modi has urged his Ministry of Water Resources to advance long-standing plans for the inter-linking of rivers, which aims to ensure water availability in all parts of the country.[104] The idea of river inter-linking goes back to Sir Arthur Cotton (best known for his successful irrigation schemes on the Kaveri, Godavari and Krishna rivers in South India), who in 1858 proposed to connect the Indian subcontinent by a grid of navigation and irrigation channels for the purpose of flood control and communication. Cotton left the country as a broken man after the triumph of the railways and the spectacular failure of several private irrigation ventures he had supported. But the idea of river inter-linking was firmly planted, and independent governments have followed it up time and again. According to critics, multi-purpose river valley development and river inter-linking present dramatic hydraulic interventions, entailing forest loss, soil degradation, waterlogging, salinisation etc., while destroying aquatic habitats and altering rivers' bio-chemical and physical properties. Existing dams' economic benefits, they claim, have drastically fallen short of expectations, while uprooting tens of thousands of people from their homes.[105] India's urban environmental policies have similarly followed colonial precedents. In what presents an almost unbroken continuity, many Indian cities for instance have forcibly relocated slum dwellers and villagers, in order to provide space for city parks and housing units for the metropolitan elite.[106]

In sum, British colonial rule has left its marks on the Indian environment, and colonial legacies continue to influence India's environmental policies up till today. It would, however, be wrong to blame India's environmental problems entirely on colonial rule. India is free to confront and break its colonial legacy, by remodelling adverse administrational and fiscal structures and devising different, more environment-friendly approaches.[107] Moreover, the country must address a number of other factors not directly related to colonial rule that negatively affect environmental politics, especially the rampant corruption and the (often connected) failure to implement existing environmental laws.[108]

It remains to be seen whether the 'Mission Clean Ganga', started in 2009 under the authority of the newly created National Ganga River Basin Authority (NGRBA), will succeed in overcoming these long-standing deficits. Recent press reports have slammed the NGRBA as 'ineffective, toothless and a non-starter', pointing out that till September 2014, it convened only three meetings without any concrete results. Several NGO representatives such as Rajendra Singh (Tarun Bharat Sangh, Jaipur) and Ravi Chopra (People's Science Institute, Dehradun) have already resigned their membership. [109] The body's latest meeting in March 2015 was left unattended by the Chief Ministers of Uttar Pradesh and West Bengal.[110] On the background of the present administrative shortcomings, the rampant corruption, and not least the strong reluctance displayed on part of the industrial community, Prime Minister Narendra Modi is indeed facing a formidable task in saving the Ganges.[111]

Notes

1 Formerly called sanitary commissioner.
2 APAC, IOR, P/10307, GoBeng, Municipal Dpt., Sanitation Branch, Prgs December 1918, Nos 3–6.
3 APAC, IOR, P/10521, GoBeng, Municipal Dpt., Sanitation Branch, Prgs May 1919 (B).
4 *Bengal Legislative Council Proceedings (Debates), 13th Session (15–21 August), Vol. XIII (1923)*, p. 362; *Bengal Legislative Council Proceedings (Debates), 19th Session (3–11 December), Vol. XIX (1925)*, pp. 280–91.
5 *Annual Report of the Director of Public Health of the United Provinces [henceforth: ARDPH UP] for the year 1926*, p. xi.
6 *ARDPH UP for the year 1928*, pp. iii–iv.
7 Ibid., p. ii.
8 'General report of the Sanitary Conference held at Nainital from 4th to 19th September 1908', APAC, IOR, V/27/840/32, 'United Provinces Sanitary Conference, Collection of papers relating to the Sanitary Conference, 1908'.
9 'Reply to Kashi Tirth Sudhar Trust', 5.1.1927, Uttar Pradesh State Archives, Regional Archives Varanasi [henceforth: UPSA(V)], Dpt. X, File no. 87, Box no. 4A, 'Kashi Sudhar Trust at Benares'.
10 Trust for the Improvement of the *tīrtha* Kashi.
11 Kashi Tirth Sudhar Trust's address to the Governor-General of India, 5.1.1927, UPSA(V), Dpt. X, File no. 87, Box no. 4A, 'Kashi Sudhar Trust at Benares'.
12 Head Assistant to the Commissioner, Benares Division, to Chairman, Benares Municipal Board, February 1928, ibid.
13 Chairman, Benares Municipal Board, to Commissioner, Benares Division, 9.7.1928, UPSA(V), Dpt. XXIII, File No. 353, Box No. 145, 'Enhancement of pilgrim tax at Benares for the purpose of protecting ghats from damage and pollution'.
14 'Note on Kashi Tirth Sudhar Trust', 4.6.1935, ibid.
15 Resolution, Benares Municipal Board, 28.4.1932, ibid.
16 Letter to the Commissioner, Benares Division, 20.7.1935, ibid.
17 UPSA(V), Dpt. XXIII, File No. 353, Box No. 145, 'Enhancement of pilgrim tax at Benares for the purpose of protecting ghats from damage and pollution'.

18 Major-General B.T. Mahon to Secy GoUP, 27.8.1909, APAC, IOR, P/8101, GoUP, Municipal Dpt., Prgs October 1909, No. 30.
19 Lieutenant-General R. Scallon to Secy GoUP, 4.7.1913, APAC, IOR, P/9173, GoUP, Municipal Dpt., Prgs October 1913, No. 5.
20 J. Campbell, Chairman, Kanpur Municipal Board, to Commissioner, Allahabad Division, 30.9.1909, APAC, IOR, P/8101, GoUP, Municipal Dpt., Prgs November 1909, No. 217(a); G.G. Sim, Chairman, Kanpur Municipal Board, to Commissioner, Allahabad Division, 8.9.1913, APAC, IOR, P/9173, GoUP, Municipal Dpt., Prgs October 1913, No. 7(a).
21 Report by W.B. Gordon, Sanitary Engineer, GoUP, APAC, IOR, P/9423, GoUP, Municipal Dpt., Prgs June 1914, No. 9(a).
22 Secy GoUP, to Commissioner, Allahabad Division, 12.6.1914, APAC, IOR, P/9423, GoUP, Municipal Dpt., Prgs June 1914, No. 12; Note by Chairman, Kanpur Municipal Board, 7.1.1914, ibid., No. 8(c); R.H. Williamson, Chairman, Kanpur Municipal Board, 2.5.1914, to Secy GoUP, 2.5.1914, ibid., No. 11.
23 A.W. Pim, Secy GoUP, to Lieutenant-General R. Scallon, 4.11.1914, APAC, IOR, P/9423, GoUP, Municipal Dpt., Prgs November 1914, No. 53.
24 *ARDPH UP for the year 1929*, p. ix; Chairman, Kanpur Municipal Board, to the Board of Public Health, 5.7.1927, APAC, IOR, P/11807, GoUP, Municipal Dpt., Prgs April 1929, No. 233(a); N. Kishore, Senior Vice-Chairman, Kanpur Municipal Board, to A. Monro, District Magistrate, Kanpur, 25.6.1927, ibid., No. 233(f).
25 Secy GoUP, to Secy Board of Public Health UP, 25.10.1928, ibid., No. 234.
26 K.S.H. Mohiuddin, Secy Board of Public Health UP, to Secy GoUP, 26.11.1928, ibid., No. 235. See also *ARDPH UP for the year 1926*, p. xi.
27 APAC, IOR, P/9423, GoUP, Municipal Dpt., Prgs 1914; APAC, IOR, P/9688, GoUP, Municipal Dpt., Prgs 1915.
28 Town Improvement Trusts were established from the late nineteenth century onwards as administrative institutions to guide the development of Indian cities (Mann 2015a: 297–8).
29 APAC, IOR, P/11807, GoUP, Municipal Dpt., Prgs 1929.
30 *ARDPH UP for the year 1930*, p. 14. See also *ARDPH for the year 1928*, p. xviii.
31 S.A. Harriss, Director of Public Health, 'A short note on the projected sullage farm at Lucknow', n.d., APAC, IOR, P/9423, GoUP, Municipal Dpt., Prgs June 1914, No. 33(a).
32 See the Director of Public Health's annual reports 1913–36.
33 *ARDPH UP for the year 1927*, p. xvi.
34 Ibid.; *ARDPH UP for the year 1928*, p. xi; *ARDPH UP for the year 1930*, p. 14. Studies by Awadhendra Sharan and others suggest that Delhi faced similar problems in operating its sullage and sewage farms at the time (Sharan 2014: 56–65; Prashad 2001: 136–47).
35 *ARDPH UP for the year 1933*, p. 26.
36 The Water Act is reproduced in Cullet and Koonan (2011: 135–56). See also Divan and Rosencranz 2002: 60–1.
37 Chadwick 1965: 120.
38 Hamlin 1988a.
39 Arnold 2006; Arnold 1996: 5–10; Driver 2004; Driver and Martins 2005.
40 Sharan 2011: 447; Metcalf 1994: 66–159.
41 See introduction, footnote 63.
42 See, for example: Kumar 2013; Baber 1996; Arnold 2000; Chakrabarti 2004; Chakrabarti 2012. Forestry, on the other hand, has been extensively covered by environmental historians.
43 As I have argued earlier, John Augustus Voelcker – whose opinion was taken during an informal interview shortly before Latham's arrival – may not be considered as a contemporary expert on river pollution (see Chapter 2).

44 Arnold 2000: 135.
45 Quot. ibid.
46 See Colvin 1894: 517.
47 Harrison 1994: 152.
48 Clemesha 1910: esp. 210–11. For Clemesha's further research see also his *The Bacteriology of Surface Waters in the Tropics* (Clemesha 1912).
49 Carkeet James 1906a; Carkeet James 1906b. See also his earlier *Oriental Drainage: A Guide to the Collection, Removal and Disposal of Sewage in Eastern Cities* (Carkeet James 1902). Carkeet James stresses the importance of sewage treatment where the sewage is ultimately disposed of in rivers (Carkeet James 1906a: 145–6).
50 Williams 1924. Williams at the time was chief engineer of the Bengal government's public health department (ibid.).
51 See, for example, Baber 1996: 137–83; Grove 1997; Bennet and Hodge 2011. Recently, environmental historians have reiterated the importance of using the British Empire 'as the scale and unit of analysis', as this brings to light the 'complex and varied intercolonial exchanges in environmental ideas, techniques and technologies' (Kumar *et al.* 2011b: 7).
52 J. Nield Cook, Health Officer, Madras Municipal Commission, 'Report on the sewage farm system in Madras (Town)', 14.4.1891, APAC, IOR, P/4112, GOI, Home Dpt., Sanitary Branch, Prgs November 1892, No. 7.
53 Carkeet James 1906a: 319–21.
54 Watson 1953.
55 Fowler 1908; Race 1918: 7.
56 MacLeod 1975: 352–4; Kumar 2013: 105.
57 NAI, Home Dpt., Sanitary Branch, Prgs September 1904, Nos 158–9.
58 APAC, IOR, P/7059, GOI, Home Dpt., Sanitary Branch, Prgs October 1905, Nos 365–84.
59 Kumar 2013: 110.
60 Gandy 2006: 17–19.
61 Mann 2015a: 301–2.
62 Baber 1996: 200–12; Mann 2004.
63 Sharan 2011: 426.
64 Khalid 2012; Masselos 1982.
65 The exception being Michael Anderson's study on air pollution in Calcutta (Anderson 1995).
66 Future studies could focus on other early industrial centres such as Kanpur or Ahmedabad for instance.
67 See Chapter 5.
68 J.A. Bourdillon, Offg. Secy GoBeng, to Secy Bengal Chamber of Commerce, 15.12.1894, APAC, IOR, P/4738, GoBeng, Municipal Dpt., Municipal Branch, Prgs March 1895, No. 25.
69 Chakrabarty 1989: 74–6, 80–1, 168–9.
70 The following account on air pollution draws entirely on Michael Anderson (1995).
71 Quot. in Anderson 1995: 308.
72 At this stage, government-owned devices (e.g. those used to drive municipal water pumps) were exempted from regulation (ibid.: 311).
73 Ibid.: 321.
74 Ibid.: 323. 1926 for example saw the construction of the Central Smoke Observatory on top of one of the highest buildings of the city, which enabled the smoke inspector to survey nearly every factory chimney within a circumference of 80 square miles (ibid.)
75 *Septic Tank Committee Report*, pp. 5–6; H.M. Haywood, Secy Bengal Chamber of Commerce, to Secy GoBeng, Municipal Dpt., 13.1.1909, APAC, IOR, P/8140, GoBeng, Municipal Dpt., Sanitation Branch, Prgs April 1909, No. 3; H.M.

Haywood, Secy Bengal Chamber of Commerce, to Secy GoBeng, Municipal Dpt., 20.3.1909, APAC, IOR, P/8140, GoBeng, Municipal Dpt., Sanitation Branch, Prgs April 1909, No. 7; C.E.A.W. Oldham, Secy GoBeng, Municipal Dpt., to Secy Bengal Chamber of Commerce, 6.4.1909, APAC, IOR, P/8140, GoBeng, Municipal Dpt., Sanitation Branch, Prgs April 1909, No. 11.

76 W.W. Clemesha, 'Report on septic tanks in Bengal', n.d., APAC, IOR, P/8418, GoBeng, Municipal Dpt., Sanitation Branch, Prgs February 1910, Nos 16–17.

77 H. Wheeler, Secy GoBeng, to Sanitary Commissioner, Bengal, 21.12.1909, APAC, IOR, P/8418, GoBeng, Municipal Dpt., Sanitation Branch, Prgs February 1910, No. 18.

78 Anderson 1995: 310, 312–15.

79 In 1928, several municipal areas still had no filtered water supplies and relied on rivers, tanks and wells (*Bengal Legislative Council Proceedings (Debates), Vol. XXX, No. 1 (1928)*, p. 190).

80 Alley 2002; Zühlke 2013.

81 B. Mittra, 'Memorandum on the Benares water-supply and sewerage scheme', 16.8.1888, in *A Collection of Miscellaneous Papers Relating to the Benares Water-Works and Drainage Projects*, p. 3, UPSA, GoNWP, Municipal Dpt., File 3/63/1890, Box 532, 'Benares Water-Supply and Drainage Project'; Jas. White, Commissioner, Banaras Municipal Board, to T.W. Holderness, Secy GoNWP, 16.10.1890, ibid., p. 78.

82 Testimony of Narendra Nath Sen, *Septic Tank Committee Report*, Appendix, p. xxxix.

83 Sengupta 2011: 268–496; Heehs 1998.

84 On other occasions, liberals and conservatives had been in strong disagreement over religious questions, e.g. in regard to the Age of Consent Bill (1891) (Sengupta 2001: 239).

85 APAC, IOR, P/6796, GoBeng, Municipal Dpt. Municipal Branch, Prgs February 1904, No. 24; *Bengalee*, 17.12.1904, p. 3.

86 Collector, Banaras Division, to E.F. Oppenheim, Commissioner, Banaras Division, 21.10.1926, UPSA(V), Dpt. X, File no. 87, Box no. 4A, 'Kashi Sudhar Trust at Benares'.

87 Memorandum by Sir Auckland Colvin, 17.9.1892, APAC, IOR, P/4062, GoNWP, Municipal Dpt., Prgs October 1892, No. 68.

88 Arnold 2013: 9.

89 Guha and Martinez-Alier 1997: ch. 1.

90 Arnold 2013: 9.

91 As highlighted throughout this book, these secular/scientific definitions of 'purity' and 'pollution' themselves were in flux and fraught with controversy during the late nineteenth and early twentieth centuries.

92 Alley 2002: 121–39.

93 Alley 2002; Zühlke 2013; also Haberman 2006.

94 Mann 2015a: 302–3.

95 'Note on Kashi Tirth Sudhar Trust', 4.6.1935, UPSA(V), Dpt. XXIII, File No. 353, Box No. 145, 'Enhancement of pilgrim tax at Benares for the purpose of protecting ghats from damage and pollution'; Letter to the Commissioner, Benares Division, 20.7.1935, ibid.

96 Mann 2015a: 303, 307.

97 Renamed National River Conservation Authority in 1995.

98 Renamed National River Conservation Directorate in 1994.

99 Zühlke 2013: 161–3, 175–7; Divan and Rosencranz 2002: 3–4.

100 Divan and Rosencranz 2002: 3–4, 210–25.

101 Alley 2002: 152.

102 Divan and Rosencranz 2002: 291–3; Guha 1990.

103 D'Souza 2008: 105–7; Klingensmith 2007; Baviskar 1995.
104 w.a. 2014.
105 D'Souza 2008: 99–107.
106 Mann 2015a: 298; Mann 2005.
107 See also Mann 2015a: 310.
108 Zühlke 2013: 394–405; Divan and Rosencranz 2002: 1–4.
109 Mishra and Aggarwal 2014.
110 w.a. 2015.
111 Chauhan 2015.

Bibliography

Archival sources

British Library, London

Asia, Pacific and Africa Collections
India Office Records.

National Archives of India, New Delhi

Government of India, Home Department (Sanitary Branch, Municipal Branch).

Uttar Pradesh State Archives, Lucknow

Government of the North-Western Provinces, Municipal Department.

Uttar Pradesh State Archives: Regional Archives, Varanasi

Dpt. X, File No. 87, Box No. 4A, 'Kashi Sudhar Trust at Benares'.
Dpt. XXIII, File No. 353, Box No. 145, 'Enhancement of pilgrim tax at Benares for the purpose of protecting ghats from damage and pollution'.

Newspapers and journals

Bengalee, Calcutta
Bharat Jiwan, Benares
British Medical Journal, London
Calcutta Review, Calcutta
Deutsche Vierteljahrsschrift für öffentliche Gesundheitspflege, Braunschweig
Englishman, Calcutta
Indian Mirror, Calcutta
Pioneer, Allahabad

Online sources

Chauhan, C. (2015). Modi's Real-Time Monitoring of Ganga Pollution Hits Roadblock. *Hindustan Times*, 22 March 2015, online, www.hindustantimes.com/india-news/modi-real-time-monitoring-of-ganga-pollution-hits-roadblock/article1-1329159.aspx (accessed 27 July 2015).

Mallet, V. (2015). The Ganges: Holy, Deadly River. *FT Magazine*, 13 February 2015, online, www.ft.com/intl/cms/s/2/dadfae24-b23e-11e4-b380-00144feab7de.html#slide0 (accessed 15 October 2015).

Ministry of Environment and Forests, Central Pollution Control Board. (2009). Ganga Water Quality Trend, December 2009, online, cpcb.nic.in/upload/NewItem_168_CPCB-Ganga_Trend Report-Final.pdf (accessed 13 September 2014).

Mishra, M. (2014). Rs 20,000 Crore Spent in 28 Years, Ganga Still a Flowing Mess. *Times of India*, 14 May 2014, online, http://timesofindia.indiatimes.com/india/Rs-20000-crore-spent-in-28-years-Ganga-still-a-flowing-mess/articleshow/35090392.cms (accessed 20 September 2014).

Mishra, P.N. and M. Aggarwal. (2014). Narendra Modi Government All Set to Revive National Ganga River Basin Authority, 3 September 2014, online, www.dnaindia.com/india/report-narendra-modi-government-all-set-to-revive-national-ganga-river-basin-authority-2015819 (accessed 13 September 2014).

National Ganga River Basin Authority (NGRBA). (2011). Environmental and Social Management Framework (ESMF), Vol I – Environmental and Social analysis, January 2011, online, www.moef.nic.in/downloads/public-information/Draft ESA Volume 1. pdf (accessed 13 September 2014).

Parsai, G. (2014). No Sewage Will Be Allowed to Drain into Ganga. *The Hindu*, 28 October 2014, online, www.thehindu.com/todays-paper/tp-national/no-sewage-will-be-allowed-to-drain-into-ganga/article6539578.ece (accessed 5 November 2014).

Pradhan, K. and J. Sriram. (2014). Call of the Sinking City. PM Narendra Modi's Pledge to Change India Faces Varanasi Test. *India Today*, 20 June 2014, online, http://india today.intoday.in/story/varanasi-ganga-narendra-modi/1/367781.html (accessed 13 September 2014).

Priyadarshini, N. (2009). Ganga River Pollution in India – A Brief Report, 8 July 2009, online, www.americanchronicle.com/articles/view/109078 (accessed 1 June 2010).

w.a. (2015). Ganga Cleaning Challenging, Need Mission-Mode Approach: PM Narendra Modi. *Economic Times*, 26 March 2015, online, http://economictimes.indiatimes.com/news/politics-and-nation/ganga-cleaning-challenging-need-mission-mode-approach-pm-narendra-modi/articleshow/46707093.cms (accessed 27 July 2015).

w.a. (2014). River-Linking to Be Reality Soon? PM Modi Wants Projects to be Taken Up Immediately. *The Economic Times*, 30 December 2014, online, http://economictimes. indiatimes.com/news/economy/agriculture/river-linking-to-be-reality-soon-pm-modi-wants-projects-to-be-taken-up-immediately/articleshow/45686760.cms (accessed 2 January 2015).

Published sources

Government documents

Annual Report on the Administration of the Bengal Presidency for 1863–64. (1865). Calcutta: O.T. Cutter, Military Orphan Press.

Annual Report of the Chemical Examiner and Bacteriologist to the North-Western Provinces and Oudh and the Central Provinces, 1894. (1894). Allahabad: Government Press.

Annual Report of the Chemical Examiner and Bacteriologist to the North-Western Provinces and Oudh and the Central Provinces, 1895. (1895). Allahabad: Government Press.

Annual Reports of the Director of Public Health of the United Provinces. (1921–47). Allahabad: Government Press.

Annual Report of the Sanitary Commission for Bengal 1864–65. [1866]. [London]: w.p.

Annual Reports of the Sanitary Commissioner of Bengal. (1868–1910). Calcutta: Government Press.

Annual Report of the Sanitary Commissioner with the Government of India, 1900. (1901). Calcutta: Central Publication Branch.

Annual Reports of the Sanitary Commissioner of the North-Western Provinces and Oudh. (1868–1901). Allahabad: Government Press.

Annual Reports of the Sanitary Commissioner of the United Provinces. (1902–20). Allahabad: Government Press.

Bengal Legislative Council Proceedings (Debates), 13th Session (15–21 August), Vol. XIII (1923). (1923). Calcutta: Government Press.

Bengal Legislative Council Proceedings (Debates), 19th Session (3–11 December), Vol. XIX (1925) (1925). Calcutta: Government Press.

Bengal Legislative Council Proceedings (Debates), Vol. XXX, No. 1 (1928). (1928). Calcutta: Government Press.

Fitzjames, F. (1880). *Preliminary Report on the Sewerage and Water Supply of the City of Benaras,* Allahabad: Government of the North-Western Provinces and Oudh.

Joshi, E.B. (1965). *Uttar Pradesh District Gazetteers. Varanasi.* Allahabad: Government of Uttar Pradesh.

Nevill, H.R. (1909a). *Benares: A Gazetteer, Being Vol. XXVI of the District Gazetteers of the United Provinces of Agra and Oudh.* Allahabad: Government Press.

Nevill, H.R. (1909b). *Saharanpur: A Gazetteer, Being Vol. II of the District Gazetteers of the United Provinces of Agra and Oudh.* Allahabad: Government Press.

Proceedings of the Sanitary Commissioner for Madras for the Year 1877. (1878). Madras: Government Press.

Proceedings of the Second All-India Sanitary Conference Held at Madras, November 11th to 16th, 1912, Vol. II, Hygiene. (1913). Simla: Government Central Branch Press.

Proceedings of the Third All-India Sanitary Conference Held at Lucknow, January 19th to 27th, 1914, Vol. V, Papers. (1914). Calcutta: Thacker, Spink & Co.

Report of the Special Committee Appointed to Examine the Working of the Septic-Tank Installations in Bengal. (1905). Calcutta: Bengal Secretariat Press.

Royal Commission on Sewage Disposal. (1915). *Final Report of the Commissioners Appointed to Inquire and Report What Methods of Treating and Disposing of Sewage [...] May Properly Be Adopted.* London: His Majesty's Stationery Office.

Laws

Indian Penal Code, 1860. (2010 [1860]). Lucknow: Eastern Book Company, 32nd edn.

Law Reports. The Public General Statutes, Passed in the Thirty-Ninth and Fortieth Years of the Reign of Her Majesty Queen Victoria, 1876, Vol. XI. (1876). London: w.p.

Books and articles

Buck, A. (1879). *A Treatise on Hygiene and Public Health*. London: Sampson Low & Co.

Carkeet James, C. (1902). *Oriental Drainage: A Guide to the Collection, Removal and Disposal of Sewage in Eastern Cities*. Bombay: Times of India Press.

Carkeet James, C. (1906a). *Drainage Problems of the East, Vol. 1*. Bombay: The Times of India Office.

Carkeet James, C. (1906b). *Drainage Problems of the East, Vol. 2*. Bombay: The Times of India Office.

Chadwick, E. (1965). *Report on the Sanitary Condition of the Labouring Population of Great Britain*, ed. by M.W. Flinn. Edinburgh: Edinburgh University Press.

Clemesha, W.W. (1910). *Sewage Disposal in the Tropics*. Calcutta, Shimla: Thacker, Spink and Co.

Clemesha, W.W. (1912). *The Bacteriology of Surface Waters in the Tropics*. Calcutta: Thacker, Spink and Co.

Clowes, F. and A.C. Houston. (1904). *The Experimental Bacterial Treatment of London Sewage: Being an Account of the Experiments Carried Out by the London County Council Between the Years 1892 and 1903*. London: London County Council.

Colvin, Sir Auckland. (1894). Municipal and Village Water Supply and Sanitation in the North-Western Provinces and Oudh. *Journal of the Society of Arts*, XLII (11 May): 515–46.

Cuningham, J. (1884). *Cholera: What Can the State do to Prevent it?*. Calcutta: Superintendent of Government Printing.

Engels, F. (2009). *The Condition of the Working Class in England*, ed. by D. McLellan. Oxford: Oxford University Press.

Fowler, G.J. (1908). Report on the Effect of the Mechanical, Chemical and Biological Purification of Sewage. The Treatment of Sewage Under Tropical Conditions. In *Bericht über den XIV. internationalen Kongress für Hygiene und Demographie. Berlin 23.-29. September 1907*, 46–63. Berlin: Verlag von August Hirschwald.

Goode, S.W. (1916). *Municipal Calcutta. Its Institutions in their Origin and Growth*. Edinburgh: T. and A. Constable.

Hankin, E.H. (1896). L'action bactéricide des eaux de la Jumna et du Ganges sur le microbe du choléra. *Annales de l'Institut Pasteur*, X: 511–23.

Hewlett, R.T. (1939). Obituary, Dr E.H. Hankin. *Nature*, 143 (29 April): 711–12, online, www.nature.com/nature/journal/v143/n3626/abs/143711b0.html (accessed 29 June 2014).

Jha, G. (trans.) (1920–6). *Manu Smṛti. The Laws of Manu*, 5 Vols. Calcutta: University of Calcutta.

Jordan, E.O. (1903). The Self-Purification of Streams. In *Investigations Representing the Departments Zoology, Anatomy, Physiology, Neurology, Botany, Pathology, Bacteriology*, ed. University of Chicago, 81–9. Chicago: The University of Chicago Press.

Latham, B. (1878). *Sanitary Engineering: A Guide to the Construction of Works of Sewerage and House Drainage*. 2nd edn, n.p.: Spon.

Latham, B. (1884). *Sanitary Engineering: A Guide to the Construction of Works of Sewerage and House Drainage*. New York: Engineering News Publishing Co.

Parkes, E. (1864). *A Manual of Practical Hygiene*. London: John Churchill & Sons.

Pettenkofer, M. v. (1890). *Die Verunreinigung der Isar durch das Schwemmsystem von München. Vortrag, gehalten im ärztlichen Verein zu München am 7. Mai 1890*. München: Rieger.

Pettenkofer, M. v. (1891). Ueber Selbstreinigung der Flüsse. *Deutsche Medizinische Wochenschrift*, 17(47): 1277–81.

Race, J. (1918). *Chlorination of Water*. New York: John Wiley & Sons.

Sedgwick, W.T. (1902). *Principles of Sanitary Science and the Public Health. With Special Reference to the Causation and Prevention of Infectious Diseases*. New York: The Macmillan Company.

Serafini, A. (1891). Contributo allo studio sperimentale dell'autodepurazione dell'acqua, specialmente dei fiumi. In *Annali dell'Instituto d'Igiene Sperimentale della R. Università di Roma*, Vol. I., 277–354. Rome: Ermanno Loescher.

Silk, A.E. (1903). *A Sewage Disposal Experiment in Calcutta*. 2nd edn, Calcutta.

Snow, J. (1849). *On the Mode of Communication of Cholera*. London: John Churchill.

Turneaure, F.E. and H.L. Russell (1901). *Public Water Supplies. Requirements, Resources, and the Construction of Works*. New York: John Wiley & Sons.

Voelcker, J.A. (1893). *Report on the Improvement of Indian Agriculture*. London: Eyre & Spottiswoode.

Watson, H.E. (1953). Gilbert John Fowler. *Journal of the Chemical Society*, 4191–2.

Williams, G.B. (1924). *Sewage Disposal in India and the East*. Calcutta, Shimla: Thacker, Spink & Co.

Wilson, G. (1873). *A Handbook of Hygiene*, London: w.p.

Wilson, H.H. (2001). *The Vishnu Purana. With an Introduction by Michael Franklin*. London: Ganesha Publishing Ltd.

w.a. (1901a). William Santo Crimp. *The Engineer*, 3 May 1901, 458.

w.a. (1901b). William Santo Crimp. In *Minutes of Proceedings, Vol. 145*, ed. Institution of Civil Engineers, 343–6.

w.a. (1937). Dr. J.A. Voelcker C.I.E. *Nature*, 140 (11 December): 1001–2.

Literature

Abraham, C.M. (1999). *Environmental Jurisprudence in India*. The Hague: Kluwer Law International.

Adas, M. (1989). *Machines as the Measure of Men: Science, Technology, and Ideologies of Western Dominance*. Ithaca: Cornell University Press.

Agarwal, D.K. *et al.* (1976). Physico-Chemical Characteristics of Ganges Water at Varanasi. *Indian Journal of Environmental Health*, 18(3): 201–6.

Ahmed, S. (1990). Cleaning the River Ganga: Rhetoric and Reality. *Ambio*, 19(1): 42–5.

Alley, K.D. (2002). *On the Banks of the Ganga. When Wastewater Meets a Sacred River*. Ann Arbor: University of Michigan Press.

Alley, K.D. (1994). Ganga and Gandagi: Interpretations of Pollution and Waste in Benaras. *Ethnology*, 33(2): 127–45.

Anderson, M.R. (1995). The Conquest of Smoke: Legislation and Pollution in Colonial Calcutta. In *Nature, Culture, Imperialism. Essays on the Environmental History of South Asia*, eds. D. Arnold and R. Guha, 293–335. New Delhi: Oxford University Press.

Arnold, D. (1985). Medical Priorities and Practice in Nineteenth-Century British India. *South Asia Research*, 5: 167–83.

Arnold, D. (1993). *Colonizing the Body. State Medicine and Epidemic Disease in Nineteenth-Century India*. Berkeley, Los Angeles: University of California Press.

Arnold, D. (1996). Introduction: Tropical Medicine before Manson. In *Warm Climates and Western Medicine: The Emergence of Tropical Medicine, 1500–1900*, ed. D. Arnold, 1–19. Amsterdam: Rodopi.

Arnold, D. (2000). *Science, Technology and Medicine in Colonial India*. Cambridge: Cambridge University Press.

Arnold, D. (2006). *The Tropics and the Traveling Gaze. India, Landscape, and Science, 1800–1856*. Seattle, London: University of Washington Press.

Arnold, D. (2010). The Ecology and Cosmology of Disease in the Banaras region. In *Culture and Power in Banaras. Community, Performance, and Environment, 1800–1980*, ed. S.B. Freitag, 246–67. 2nd edn, New Delhi: Oxford University Press.

Arnold, D. (2013). Pollution, Toxicity and Public Health in Metropolitan India, 1850–1939. *Journal of Historical Geography*, 42: 124–33.

Arnold, D. and E. DeWald. (2011). Cycles of Empowerment? The Bicycle and Everyday Technology in Colonial India and Vietnam. *Comparative Studies in Society and History*, 53(4): 971–96.

Baber, Z. (1996). *The Science of Empire. Scientific Knowledge, Civilization, and Colonial Rule in India*. Albany: State University of New York Press.

Baird, R.D. ed. (2001). *Religion in Modern India*. 4th rev. edn, New Delhi: Manohar.

Barles, S. (2005). *L'invention des déchets urbains: France 1790–1970*. Seyssel: Champ Vallon.

Basu, S. (1995). Emergence of the Mill Towns in Bengal 1880–1920: Migration Pattern and Survival Strategies of Industrial Workers. *The Calcutta Historical Journal*, 18: 97–134.

Baviskar, A. (1995). *In the Belly of the River: Tribal Conflicts over Development in the Narmada Valley*. Delhi: Oxford University Press.

Bayly, C. (1983). *Rulers, Townsmen, and Bazaars: North Indian Society in the Age of British Expansion, 1770–1870*. Cambridge: Cambridge University Press.

Beattie, J. (2011). *Empire and Environmental Anxiety. Health, Science, Art and Conservation in South Asia and Australasia, 1800–1920*. Houndmills: Palgrave Macmillan.

Beder, S. (1993). From Sewage Farms to Septic Tanks: Trials and Tribulations in Sidney. *Journal of the Royal Australian Historical Society*, 79(1/2): 72–95, online, www.uow.edu.au/~sharonb/sewage/history2.html (accessed 14 July 2014).

Beinart, W. and L. Hughes. eds. (2007). *Environment and Empire*. New York: Oxford University Press.

Bellwinkel-Schempp, M. (1982). Kanpur 1830–1973: Eine koloniale Industriestadt und ihre Arbeiterschaft. In *Städte in Südasien. Geschichte, Gesellschaft, Gestalt*, eds H. Kulke *et al.*, 133–57. Wiesbaden: Franz Steiner Verlag.

Benidickson, J. (2007). *The Culture of Flushing. A Social and Legal History of Sewage*. Vancouver, Toronto: UBC Press.

Bennet, B.M. and J.M. Hodge. eds. (2011). *Science and Empire. Knowledge and Networks of Science Across the British Empire, 1800–1970*. Basingstoke: Palgrave Macmillan.

Biswas, A.B. (1990). Water supply in Calcutta. In *Calcutta. The Living City. Vol. II: The Present*, ed. S. Chaudhuri, 160–5. Calcutta: Orient Longman.

Boehmer, E. (2002). *Empire, the National, and the Postcolonial 1890–1920*. Oxford: Oxford University Press.

Breeze, L.E. (1993). *The British Experience with River Pollution, 1865–1876*. New York etc.: Peter Lang.

Brockington, J. (2008). Dharmaśāstras. In *Encyclopedia of Hinduism*, eds D. Cush *et al.*, 186–8. London: Routledge.

Broich, J. (2007). Engineering the Empire: British Water Supply Systems and Colonial Societies, 1850–1900. *Journal of British Studies*, 46(2): 346–65.

Brundage, A. (1988). *England's 'Prussian Minister': Edwin Chadwick and the Politics of Government Growth, 1832–1854*. Pennsylvania: Pennsylvania State University Press.

Büschenfeld, J. (1997). *Flüsse und Kloaken. Umweltfragen im Zeitalter der Industrialisierung (1870–1918)*. Stuttgart: Klett-Cotta.

Centre for Science and Environment (2007). *Sewage Canal. How to Clean the Yamuna*. New Delhi: Centre for Science and Environment.

Chakrabarti, P. (2004). *Western Science in Modern India. Metropolitan Methods, Colonial Practices*. Delhi: Permanent Black.

Chakrabarti, P. (2012). *Bacteriology in British India. Laboratory Medicine and the Tropics*. Rochester: University of Rochester Press.

Chakrabarti, P. (2014). *Medicine and Empire 1600–1960*. Basingstoke: Palgrave Macmillan.

Chakrabarti, P. (2015). Purifying the river: Pollution and Purity of Water in Colonial Calcutta. *Studies in History*, 31(2): 178–205.

Chakrabarti, R. ed. (2007). *Situating Environmental History*. New Delhi: Manohar.

Chakrabarty, D. (1989). *Rethinking Working-Class History: Bengal 1890–1940*. Princeton: Princeton University Press.

Chakraborty, R.N. *et al.* (1965). Stream Pollution and its Effect on Water Supply. A Report of Survey. In *Problems in Water Treatment: Symposia Proceedings*, ed. R.S. Mehta, 211–19. Nagpur: Central Public Health Engineering Research Institute.

Chapman, G.P. (1995). The Ganges and Brahmaputra Basins. In *Water and the Quest for Sustainable Development in the Ganges Valley*, eds G.P. Chapman and M. Thompson, 3–24. London, New York: Mansell.

Chapple, C.K. and M.E. Tucker. eds. (2000). *Hinduism and Ecology: The Intersection of Earth, Sky, and Water*. Cambridge, MA: Harvard University Center for the Study of World Religions.

Choudhury, D.K.L. (2010). *Telegraphic Imperialism: Crisis and Panic in the Indian Empire, c.1830–1920*. Basingstoke: Palgrave Macmillan.

Cronon, W. (1991). *Nature's Metropolis. Chicago and the Great West*. New York, London: Norton.

Cullet, P. and J. Gupta. (2009). India: Evolution of Water Law and Policy. In *The Evolution of the Law and Politics of Water*, eds J.W. Dellapenna and J. Gupta, 157–73. [Dordrecht]: Springer Science + Business Media B.V.

Cullet, P. and S. Koonan. eds. (2011). *Water Law in India. An Introduction to Legal Instruments*. New Delhi: Oxford University Press.

Curtin, P.D. (1989). *Death by Migration. Europe's Encounter with the Tropical World in the Nineteenth Century*. Cambridge: Cambridge University Press.

D'Orazio, U. (1998). Scienza tedesca e università italiana: Recezione di modelli esteri nell'istituzionalizzazione delle discipline igieniche in Italia (1885–1900). *Medizinhistorisches Journal*, 33(3/4): 293–321.

D'Souza, R. (2006). Water in British India: The Making of a 'Colonial Hydrology'. *History Compass*, 4(4): 621–8.

D'Souza, R. (2008). River-Linking and its Discontents. In *Water First. Issues and Challenges for Nations and Communities in South Asia*, eds K. Lahiri-Dutt and R.J. Wasson, 99–121. New Delhi: Sage.

Danino, M. (2010). *The Lost River. On the Trail of the Sarasvati*. New Delhi etc.: Penguin.

Darian, S.G. (2001). *The Ganges in Myth and History*. 1st Indian edn, Delhi: Motilal Banarsidass.

Das, A. (2007). Industrial Workers, their Environmental Crises and Empire: Perspectives on Colonial Bengal. In *Situating Environmental History*, ed. R. Chakrabarti, 243–61. New Delhi: Manohar.

Das Gupta, S.P. ed. (1984). *Basin, Sub-Basin Inventory of Water Pollution: The Ganga Basin, Part 2*. New Delhi: Central Board for the Prevention and Control of Water Pollution.

Das Gupta, S.P. (1990). The Site of Calcutta. Geology and Physiography. In *Calcutta. The Living City. Volume I: The Past*, ed. S. Chaudhuri, 2–4. Calcutta: Oxford University Press.

Datta, P. (2009). Ranald Martin's *Medical Topography* (1837): The Emergence of Public Health in Calcutta. In *The Social History of Health and Medicine in Colonial India*, eds B. Pati and M. Harrison, 15–30. London: Routledge.

Datta Munshi, J.S. (2012). *The River Ganga: The Life Line of India*. New Delhi: Daya Publishing House.

Divan, S. and A. Rosencranz (2002). *Environmental Law and Policy in India. Cases, Materials and Statutes*. 2nd edn, New Delhi: Oxford University Press.

Divyabhanusinh (1995). *The End of a Trail: The Cheetah in India*. New Delhi: Banyan Books.

Divyabhanusinh (2005). *The Story of Asia's Last Lions*. Mumbai: Marg Publications.

Dossal, M. (1991). *Imperial Designs and Indian Realities: The Planning of Bombay City, 1845–1875*. Bombay: Oxford University Press.

Douglas, M. (2002). *Purity and Danger. An Analysis of Concepts of Pollution and Taboo*. Routledge Classic Edition, London, New York: Routledge.

Drayton, R. (2000). *Nature's Government. Science, Imperial Britain, and the 'Improvement' of the World*. New Haven, London: Yale University Press.

Driver, F. (2004). Imagining the Tropics: Views and Visions of the Tropical World. *Singapore Journal of Tropical Geography*, 25(1): 1–17.

Driver, F. and L. Martins (2005). Views and Visions of the Tropical World. In *Tropical Visions in an Age of Empire*, eds F. Driver and L. Martins, 3–20. Chicago: Chicago University Press.

Eck, D.L. (1993). *Banaras. City of Light*. New Delhi etc.: Penguin.

Eck, D.L. (1996). Ganga. The Goddess Ganges in Hindu Sacred Geography. In *Devi. Goddesses of India*, eds J.S. Hawley and D.M. Wulff, 137–53. Berkeley etc.: University of California Press.

Freitag, S.B. (2010a). Introduction: The History and Political Economy of Banaras. In *Culture and Power in Banaras. Community, Performance, and Environment, 1800–1980*, ed. S.B. Freitag, 1–22. 2nd edn, New Delhi: Oxford University Press.

Freitag, S.B. (2010b). State and Community: Symbolic Popular Protest in Banaras's Public Arenas. In *Culture and Power in Banaras. Community, Performance, and Environment, 1800–1980*, 203–28. 2nd edn, New Delhi: Oxford University Press.

Furedy, C. (1987). From Waste Land to Waste-Not Land: The Role of the Salt Lakes, East Calcutta, in Waste Treatment and Recycling, 1845–1930. In *Calcutta: The Urban Experience*, ed. P. Sinha, 145–53. Calcutta: Riddhi-India.

Gadgil, M. and R. Guha (1992). *This Fissured Land. An Ecological History of India*. New Delhi: Oxford University Press.

Gadgil, M. and R. Guha (2007). Ecological Conflicts and the Environmental Movement in India. In *Environmental Issues in India. A Reader*, ed. M. Rangarajan, 385–428. New Delhi: Dorling Kindersley.

Gandy, M. (2006). The Bacteriological City and its Discontents. *Historical Geography*, 34: 14–25.

Gandy, M. (2008). Landscapes of Disaster: Water, Modernity, and Urban Fragmentation in Mumbai. *Environment and Planning A*, 40: 108–30.

Ghose, S. (1993). *Jawaharlal Nehru. A Biography*. New Delhi: Allied Publishers Ltd.

Gillion, K.L. (1968). *Ahmedabad: A Study in Indian Urban History*. Berkeley: University of California Press.

Goddard, N. (1996). 'A Mine of Wealth'? The Victorians and the Agricultural Value of Sewage. *Journal of Historical Geography*, 22(3): 274–90.

Goddard, N. (2004). Voelcker, (John Christopher) Augustus. In *Oxford Dictionary of National Biography*, Oxford: Oxford University Press, online, www.oxforddnb.com/view/article/28345 (accessed 25 June 2013).

Godley, A. (2001). The Global Diffusion of the Sewing Machine, 1850–1914. *Research in Economic History*, 20: 1–45.

Goswami, M. (2004). *Producing India: From Colonial Economy to National Space*. Chicago: University of Chicago Press.

Grove, R. (1997). *Green Imperialism. Colonial Expansion, Tropical Island Edens and the Origins of Environmentalism, 1600–1860*. Repr., Cambridge: Cambridge University Press.

Guha, R. (1989). *The Unquiet Woods: Ecological Change and Peasant Resistance in the Himalaya*. New Delhi: Oxford University Press.

Guha, R. (1990). An Early Environmental Debate: The Making of the 1878 Forest Act. *Indian Economic and Social History Review*, 27(1): 67–84.

Guha, R. (2000). *Environmentalism. A Global History*. New York: Longman.

Guha, R. and J. Martinez-Alier (1997). *Varieties of Environmentalism: Essays North and South*. London: Earthscan.

Gupta, A.J. (1997). Defying Death: Nationalist Revolutionism in India 1897–1938. *Social Scientist*, 25(9/10): 3–27.

Haberman, D.L. (2006). *River of Love in an Age of Pollution. The Yamuna River of Northern India*. Berkeley etc.: University of California Press.

Halliday, S. (2001). *The Great Stink of London. Sir Joseph Bazalgette and the Cleansing of the Victorian Metropolis*. Stroud: Sutton Publishing.

Halliday, S. (2007). *The Great Filth. The War Against Disease in Victorian England*. Stroud: Sutton Publishing.

Hamlin, C. (1987). *What Becomes of Pollution? Adversary science and the controversy on the self-purification of rivers in Britain, 1850–1900*. New York, London: Garland Publishing Inc.

Hamlin, C. (1988a). Politics and germ theories in Victorian Britain: The Metropolitan Water Commissions of 1867–9 and 1892–3. In *Government and Expertise. Specialists, Administrators and Professionals, 1860–1919*, ed. R. MacLeod. Cambridge etc.: Cambridge University Press.

Hamlin, C. (1988b). William Dibdin and the Idea of Biological Sewage Treatment. *Technology and Culture*, 29(2): 189–218.

Hamlin, C. (1988c). Muddling in Bumbledom: On the Enormity of Large Sanitary Improvements in Four British towns, 1855–1885. *Victorian Studies*, 32(1): 55–83.

Hamlin, C. (1990). *A Science of Impurity. Water Analysis in Nineteenth Century Britain*. Bristol: Adam Hilger.

Hardy, A. (2001). *Health and Medicine in Britain Since 1860*. Houndmills: Palgrave.

Hardy, A. (2005). *Ärzte, Ingenieure und städtische Gesundheit. Medizinische Theorien in der Hygienebewegung des 19. Jahrhunderts*. Frankfurt: Campus.

Harrison, J.B. (1980). Allahabad: A Sanitary History. In *The City in South Asia. Pre-Modern and Modern*, eds K. Ballhatchet and J. Harrison, 167–95. London: Curzon Press.

Harrison, M. (1994). *Public Health in British India. Anglo-Indian Preventive Medicine 1859–1914*. Cambridge: Cambridge University Press.

Harrison, M. (1996). A Question of Locality: The Identity of Cholera in British India, 1860–1890. In *Warm Climates and Western Medicine: The Emergence of Tropical Medicine, 1500–1900*, ed. D. Arnold, 133–59. Amsterdam: Rodopi.

Harrison, M. (2011). Medicine and Colonialism in South Asia Since 1500. In *The Oxford Handbook of the History of Medicine*, ed. M. Jackson, 285–301. New York: Oxford University Press.

Hatcher, B.A. (2004). Contemporary Hindu Thought. In *Contemporary Hinduism. Ritual, Culture, and Practice*, ed. R. Rinehart, 179–211. Santa Barbara: ABC-CLIO.

Hauser, B. (2011). Das Vermitteln der Regel(n): Menstruelle Unreinheit in der performativen Praxis indischer Frauen. In *Reinheit*, eds P. Burschel and C. Marx, 195–215. Wien etc.: Böhlau Verlag.

Hayavadana Rao, C. [1916]. *The Indian Biographical Dictionary, 1915*. Madras: w.p.

Hays, S.P. (1998). The Role of Urbanization in Environmental History. In *Explorations in Environmental History: Essays*, ed. S.P. Hays, 69–100. Pittsburgh: University of Pittsburgh Press.

Headrick, D. (1981). *The Tools of Empire: Technology and European Imperialism in the Nineteenth Century*. New York: Oxford University Press.

Headrick, D. (1988). *The Tentacles of Progress. Technology Transfer in the Age of Imperialism, 1850–1940*. New York: Oxford University Press.

Heehs, P. (1998). *Nationalism, Terrorism, Communalism. Essays in Modern Indian History*. New Delhi: Oxford University Press.

Heehs, P. (2010). Revolutionary Terrorism in British Bengal. In *Terror and the Postcolonial*, eds E. Boehmer and S. Morton, 153–76. Chichester: Wiley-Blackwell.

Heitler, R. (1972). The Varanasi House Tax Hartal of 1810–11. *Indian Economic and Social History Review*, 9: 239–57.

Huber, V. (2006). The Unification of the Globe by Disease? The International Sanitary Conferences on Cholera, 1851–1894. *The Historical Journal*, 49(2): 453–76.

Hust, E. (2005). Introduction: Problems of Urbanization and Urban Governance in India. In *Urbanization and Governance in India*, eds E. Hust and M. Mann, 1–26. New Delhi: Manohar.

Insley, J. (2004). Latham, Baldwin (1836–1917). In *Oxford Dictionary of National Biography*, online, www.oxforddnb.com/templates/article.jsp?articleid=52328&back (accessed 11 February 2014).

Isaacs, J.D. (1998). D D Cunningham and the Aetiology of Cholera in British India, 1869–1897. *Medical History*, 42: 279–305.

Jain, S. (2011). *In Search of Yamuna. Reflections on a River Lost*. New Delhi: Vitasta Publishing.

Jones, E.M. (2013). *Parched City. A History of London's Public and Private Drinking Water*. Winchester: Zero Books.

Joshi, A. (2008). *Town Planning: Regeneration of Cities*. New Delhi: New India Publishing Agency.

Khalid, A. (2012). 'Unscientific and Insanitary'. Hereditary Sweepers and Customary Rights in the United Provinces. In *Public Health in the British Empire: Intermediaries, Subordinates and Public Health Practice, 1850–1960*, eds R. Johnson and A. Khalid, 51–70. New York: Routledge.

Kidambi, P. (2007). *The Making of an Indian Metropolis. Colonial Governance and Public Culture in Bombay, 1890–1920*. Ashgate: Aldershot.

King, A. (1980). Colonialism and the Development of the Modern South Asian City: Some Theoretical Considerations. In *The City in South Asia: Pre-Modern and Modern*, ed. K. Ballhatchet, 1–19. London: Curzon Press.

King, A.S. (2005). The Ganga: Waters of Devotion. In *The Intimate Other: Love Divine in Indic Religions*, eds A.S. King and J.L. Brockington, 153–93. New Delhi: Orient Longman.

Kinsley, D. (2005). *Hindu Goddesses. Vision of the Divine Feminine in the Hindu Religious Tradition*. Repr., Delhi: Motilal Banarsidass.

Klein, I. (1980). Cholera: Theory and Treatment in Nineteenth Century India. *Journal of Indian History*, 58: 35–51.

Klein, I. (1994). Imperialism, Ecology and Disease: Cholera in India, 1850–1950. *Indian Economic and Social History Review*, 31(4): 491–518.

Klingensmith, D. (2007). *'One Valley and a Thousand'. Dams, Nationalism, and Development*. New Delhi: Oxford University Press.

Kumar, A. (1998). *Medicine and the Raj. British Medical Policy in India, 1835–1911*. New Delhi: Sage.

Kumar, D. ed. (2001). *Disease & Medicine in India. A Historical Overview*. New Delhi: Tulika.

Kumar, D. (2013). *Science and the Raj*. 2nd edn, repr., New Delhi: Oxford University Press.

Kumar, D. *et al.* eds (2011a). *The British Empire and the Natural World. Environmental Encounters in South Asia*. New Delhi: Oxford University Press.

Kumar, D. *et al.* eds (2011b). Introduction. In *The British Empire and the Natural World. Environmental Encounters in South Asia*, eds D. Kumar et al., 1–13. New Delhi: Oxford University Press.

Kumar, N. (2010). Work and Leisure in the Formation of Identity: Muslim Weavers in a Hindu City. In *Culture and Power in Banaras. Community, Performance, and Environment, 1800–1980*, ed. S.B. Freitag, 147–70. 2nd edn, New Delhi: Oxford University Press.

Luckin, B. (1986). *Pollution and Control. A Social History of the Thames in the Nineteenth Century*. Bristol, Boston: Adam Hilger.

MacLeod, R. (1975). Scientific advice for British India: Imperial Perceptions and Administrative Goals, 1898–1923. *Modern Asian Studies*, 9(3): 343–84.

MacLeod, R. and D. Kumar, eds (1995). *Technology and the Raj: Western Technology and Technical Transfers to India, 1700–1947*. New Delhi: Sage.

Majumdar, R.C. (1981). *History of Modern Bengal. Part Two (1905–1947)*. Calcutta: G. Bharadwaj & Co.

Mann, M. (2004). 'Torchbearers Upon the Path of Progress': Britain's Ideology of a 'Moral and Material Progress' in India. In *Colonialism as Civilizing Mission: Cultural Ideology in British India*, eds M. Mann and H. Fischer-Tiné, 1–26. London: Anthem Press.

Mann, M. (2005). Town Planning and Urban Resistance in the Old City of Delhi, 1937–77. In *Urbanization and Governance in India*, eds E. Hust and M. Mann, 251–78. New Delhi: Manohar.

Mann, M. (2007). Delhi's Belly: On the Management of Water, Sewage and Excreta in a Changing Urban Environment During the Nineteenth Century. *Studies in History*, 23(1): 1–31.

Mann, M. (2013). Environmental History and Historiography on South Asia: Context and Some Recent Publications. *Suedasien-Chronik/South Asia Chronicle*, 3: 324–57, online, http://edoc.hu-berlin.de/docviews/abstract.php?lang=ger&id=40590 (accessed 27 May 2014).

Mann, M. (2015a). *South Asia's Modern History. Thematic Perspectives*. London: Routledge.

Mann, M. (2015b). Cholera in den Zeiten der Globalisierung. Oder wie die Welt in zwei Teile zerfällt. In *Europa jenseits der Grenzen. Festschrift für Reinhard Wendt*, eds M. Mann and J.G. Nagel, 389–431. Heidelberg: Draupadi Verlag.

Markandya, A. and M.N. Murty (2000). *Cleaning-Up the Ganges. A Cost-Benefit Analysis of the Ganga Action Plan*. New Delhi: Oxford University Press.

Masselos, J. (1982). Jobs and Jobbery: The Sweeper in Bombay under the Raj. *Indian Economic and Social History Review*, 19(2): 101–39.

Mathur, L.N. (1980) A Federal Legislative History of Control of Water Pollution in India. In *Legal Control of Environmental Pollution*, ed. S.L. Agarwal, 86–94. Bombay: N.M. Tripathi.

Meckel, R.A. (2001). *Save the Babies: American Public Health Reform and the Prevention of Infant Mortality, 1850–1929*. Michigan: University of Michigan Press.

Meisner Rosen, C. and J.A. Tarr (1994). The Importance of an Urban Perspective in Environmental History. *Journal of Urban History*, 20(3): 299–310.

Melosi, M.V. (1993). The Place of the City in Environmental History. *Environmental History Review*, 17(1): 1–23.

Melosi, M.V. (2001). *Effluent America. Cities, Industry, Energy, and the Environment*. Pittsburgh: University of Pittsburgh Press.

Melosi, M.V. (2008). *The Sanitary City. Environmental Services in Urban America from Colonial Times to the Present. Abridged Version*. Pittsburgh: University of Pittsburgh Press.

Metcalf, T.R. (1994). *Ideologies of the Raj*. Cambridge: Cambridge University Press.

Mohajeri, S. (2005). *100 Jahre Berliner Wasserversorgung und Abwasserentsorgung 1840–1940*. Stuttgart: Franz Steiner Verlag.

Mukharji, P.B. (2011). *Nationalizing the Body: The Medical Market, Print and Daktari Medicine*. London: Anthem.

Mukherjee, N. (1968). *The Port of Calcutta*. Calcutta: Commissioners for the Port of Calcutta.

Muraleedharan, V.R. and D. Veeraraghavan (1995). Disease, Death and Local Administration: Madras City in Early 1900s. *Radical Journal of Health*, 1: 9–24.

Mushtaq, M.U. (2009). Public Health in British India: A Brief Account of the History of Medical Services and Disease Prevention in Colonial India. *Indian Journal of Community Medicine*, 34(1): 6–14.

Nair, P.T. (1990). The Growth and Development of Old Calcutta. In *Calcutta. The Living City. Vol. I: The Past*, ed. S. Chaudhuri, 10–23. Calcutta: Oxford University Press.

Nath, K.J. and A. Majumdar (1990). Drainage, Sewerage and Waste Disposal. In *Calcutta. The Living City. Volume II. The Present*, ed. S. Chaudhuri, 167–72. Calcutta: Oxford University Press.

Nejad, R.M. (2015). Religious Processions as a Means of Social Conciliation. In *Community-Based Urban Violence Prevention. Innovative Approaches in Africa, Latin America, Asia and the Arab Region*, eds K. Mathey and Silvia Mathuk, 268–79. Bielefeld: transcript.

Nelson, L.E. ed. (1998). *Purifying the Earthly Body of God: Religion and Ecology in Hindu India*. Albany: State University of New York Press.

Nelson, L.E. (2007). Shraddha. In *An Introductory Dictionary of Theology and Religious Studies*, eds O.O. Espín and J.B. Nickoloff, 1279. Collegeville: Liturgical Press.

Neri Serneri, S. (2002). Water Pollution in Italy: The Failure of the Hygienic Approach, 1890s-1960s. In *Le démon moderne. La pollution dans les sociétiés urbaines et industrielles d'Europe*, eds C. Bernhardt and G. Massard-Guilbaud, 157–78. Clermont-Ferrand: Presses Universitaires Blaise-Pascal.

Ogawa, M. (2000). Uneasy Bedfellows: Science and Politics in the Refutation of Koch's Bacterial Theory of Cholera. *Bulletin of the History of Medicine*, 74: 671–707.

Oldenburg, V. (1988). *The Making of Colonial Lucknow, 1856–1877*. Princeton: Princeton University Press.

Oosthoek, J. (2002). The Stench of Prosperity. Water Pollution in the Northern Netherlands 1850–1980. *Le démon moderne. La pollution dans les sociétiés urbaines et industrielles d'Europe*, eds C. Bernhardt and G. Massard-Guilbaud, 179–94. Clermont-Ferrand: Presses Universitaires Blaise-Pascal.

Palmer, R. (1973). *The Water Closet. A New History*. David & Charles: Newton Abbot.

Pati, B. and M. Harrison, eds. (2006). *Health, Medicine and Empire. Perspectives on Colonial India*. Repr., New Delhi: Orient Longman.

Pati, B. and M. Harrison, eds. (2009a). *The Social History of Health and Medicine in Colonial India*. London: Routledge.

Pati, B. and M. Harrison (2009b). Social History of Health and Medicine: Colonial India. In *The Social History of Health and Medicine in Colonial India*, eds B. Pati and M. Harrison, 1–14. London: Routledge.

Peers, D.M. (2012). State, Power, and Colonialism. In *India and the British Empire*, eds D.M. Peers and N. Gooptu, 16–43. Oxford: Oxford University Press.

Prashad, V. (2001). The Technology of Sanitation in Colonial Delhi. *Modern Asian Studies*, 35(1): 113–55.

Raina, D. and S.I. Habib. (1995). Bhadralok Perceptions of Science, Technology and Cultural Nationalism. *Indian Economic and Social History Review*, 32(1): 95–117.

Raj, K. (2007). *Relocating Modern Science. Circulation and the Construction of Knowledge in South Asia and Europe, 1650–1900*. Houndmills, New York: Palgrave Macmillan.

Rajan, S.R. (2006). *Modernizing Nature. Forestry and Imperial Eco-Development 1800–1950*. Oxford: Oxford University Press.

Ramanna, M. (1996). Randchodlal Chotalal: Pioneer of Public Health in Ahmedabad. *Radical Journal of Health*, 2: 99–111.

Ramanna, M. (2002). *Western Medicine and Public Health in Colonial Bombay 1845–1895*. New Delhi: Orient Longman.

Ramasubban, R. (1982). *Public Health and Medical Research in India: Their Origins and Development Under the Impact of British Colonial Policy*. Stockholm: Swedish Agency for Research Cooperation with Developing Countries.

Rangarajan, M. (1996). *Fencing the Forest. Conservation and Ecological Change in India's Central Provinces 1860–1914*. New Delhi: Oxford University Press.

Rangarajan, M. (1998). The Raj and the Natural world. The Campaign against 'Dangerous Beasts' in Colonial India. *Studies in History*, 14: 265–99.

Rangarajan, M. (2007a). *Environmental Issues in India. A Reader*. New Delhi: Dorling Kindersley.

Rangarajan, M. (2007b). Introduction. In *Environmental Issues in India. A Reader*, ed. M. Rangarajan, xiii–xxvii. New Delhi: Dorling Kindersley.

Rangarajan, M. (2008). *India's Wildlife History*. 2nd edn, Delhi: Permanent Black.

Rangarajan, M. (2009). Environmental Histories of India. Of States, Landscapes, and Ecologies. In *The Environment and World History*, eds E. Burke III and K. Pommeranz, 229–54. Berkeley etc.: University of California Press.

Rangarajan, M. (2012). Environment and Ecology under British Rule. In *India and the British Empire*, eds D.M. Peers and N. Gooptu, 212–29. Oxford: Oxford University Press.

Ray, R.K. (1988). Moderates, Extremists, and Revolutionaries: Bengal, 1900–1908. In *Congress and Indian Nationalism: The Pre-Independence Phase*, eds R. Sisson and S. Wolpert, 62–89. Berkeley etc.: University of California Press.

Reeves, P. (1995). Inland Waters and Freshwater Fisheries: Issues of Control, Access and Conservation in Colonial India. In *Nature, Culture, Imperialism. Essays on the Environmental History of South Asia*, eds D. Arnold and R. Guha, 260–92. New Delhi: Oxford University Press.

Reid, D. (1991). *Paris Sewers and Sewermen. Realities and Representations*. Cambridge MA: Harvard University Press.

Rosen, G. (1976). Edmund A. Parkes in the Development of Hygiene. *Journal of the Royal Army Medical Corps*, 122(4): 187–91.

Rosenthal, L. (2014). *The River Pollution Dilemma in Victorian England. Nuisance Law Versus Economic Efficiency*. Farnham: Ashgate.

Rothermund, D. (1988). *An Economic History of India. From Pre-Colonial Times to 1986*. London etc.: Croom Helm.

Russell, C.A. (1996). *Edward Frankland. Chemistry, Controversy and Conspiracy in Victorian England*. Cambridge etc.: Cambridge University Press.

Sampat, P. (1996). The Ganges: Myth and Reality. *World Watch*, 9(4): 24–32.

Sarkar, S. (1973). *The Swadeshi Movement in Bengal 1903–1908*. New Delhi: People's Publishing House.

Sau, S. (1997). *Port and Development: A Study of Calcutta Port in India*. Calcutta: Firma KLM.

Saxena, K.L. *et al.* (1966). Pollution Studies of the River Ganga near Kanpur. *Indian Journal of Environmental Health*, 8: 270–85.

Schmiedchen, A. (2008). Manu. In *Encyclopedia of Hinduism*, eds D. Cush *et al.*, 491. London: Routledge.

Schneider, D. (2011). *Hybrid Nature. Sewage Treatment and the Contradictions of the Industrial Ecosystem*. Cambridge MA: MIT Press.

Schott, D. (2005). Resources of the City: Towards a European Urban Environmental History. In *Resources of the City. Contributions to an Environmental History of Modern Europe*, eds D. Schott *et al.* 1–27. Aldershot: Ashgate.

Seghal, S. (1999). *Encyclopaedia of Hinduism, Vol. 2*, New Delhi: Sarup & Sons.

Sengar, D.S. (2007). *Environmental Law*, New Delhi: PHI Learning.

Sengupta, N. (2001). *History of the Bengali-speaking People*. New Delhi: UBSPD.

Sengupta, N. (2011). *Land of Two Rivers. A History of Bengal from the Mahabharata to Mujib*. New Delhi: Penguin.

Sethia, T. (1996). The rise of the Jute Manufacturing Industry in Colonial India: A Global Perspective. *Journal of World History*, 7(1): 71–99.

Sharan, A. (2011). From Source to Sink: 'Official' and 'Improved' Water in Delhi, 1868–1956. *Indian Economic and Social History Review*, 48(3): 425–62.

Sharan, A. (2014). *In the City, Out of Place. Nuisance, Pollution, and Dwelling in Delhi, c.1850–2000*. New Delhi: Oxford University Press.

Sharma, M. (2007). Polluted River or Goddess and Saviour? In *Five Emus to the King of Siam. Environment and Empire*, ed. H. Tiffin, 31–50. Amsterdam and New York: Rodopi.

Sharma, P.D. (2009). *Ecology and Environment*. Repr., New Delhi: Rastogi Publications.

Shukla, A.C. and A. Vandana. (1995). *Ganga: A Water Marvel*. New Delhi: Ashish Publishing House.

Sidwick, J.M and J.E. Murray (1976). A Brief History of Sewage Treatment. 2. The Royal Commission. *Effluent and Water Treatment Journal*, 16(4): 193–9.

Singh, D.K. (2001). 'Clouds of cholera' and clouds around cholera, 1817–1870. In *Disease & Medicine in India. A Historical Overview*, ed. D. Kumar, 144–65. New Delhi: Tulika Books.

Singh, N., ed. (2000). *Encyclopaedia of the Indian Biography. Vol. VII. S*. New Delhi: A.P.H. Publishing Corporation.

Sinha, N. (1968b). Surendranath Banerjea. In *Freedom Movement in Bengal 1818–1904. Who's Who*, ed. N. Sinha, 371–9. Calcutta: Government of West Bengal.

Sinha, N. (1968a). Narendranath Sen. In *Freedom Movement in Bengal 1818–1904. Who's Who*, ed. N. Sinha, 292–5. Calcutta: Government of West Bengal.

Sinha, P. (1990). Calcutta and the Currents of History 1690–1912. In *Calcutta. The Living City. Volume I. The Past*, ed. S. Chaudhuri, 31–44. Calcutta: Oxford University Press.

Sinha, R.K. and A. Ghosh (2008). Cleaner Production Policy and Philosophy: The Preventive Strategy of Industrial Waste Management. Some Experiences from India and Australia. In *Progress in Waste Management Research*, eds J.I. Daven and R.N. Klein, 93–158. New York: Nova Science Publishers.

Spellman, F.R. (1996). *Stream Ecology and Self-Purification. An Introduction for Wastewater and Water Specialists*. Lancaster: Technomic Publishing Company.

Srivastava, M.L. (2006). *Ganga Pollution*. New Delhi: Bookwell.

Stark, U. (2012). Knowledge in Context: Raja Shivaprasad as Hybrid Intellectual and People's Educator. In *Trans-Colonial Modernities in South Asia*, eds M.S. Dodson and B.A. Hatcher, 68–91. Abingdon: Routledge.

Steinberg, T. (1991). *Nature Incorporated. Industrialization and the Waters of New England*. Cambridge: Cambridge University Press.

Strang, V. (2006). *The Meaning of Water*. Repr., Oxford, New York: Berg.

Tarr, J.A. (1996). *The Search for the Ultimate Sink. Urban Pollution in Historical Perspective*. Akron: University of Akron Press.

Tarr, J.A. (2001). Urban History and Environmental History in the United States: Complementary and Overlapping Fields. In *Environmental Problems in European Cities of the 19th and 20th Century*, ed. C. Bernhardt, 25–39. Münster: Waxmann.

Tarr, J.A. and J.K. Stine, eds (1994). Industry, Pollution, and the Environment. *Environmental History Review* (special issue), 18.

Tejani, S. (2008). *Indian Secularism. A Social and Intellectual History, 1890–1950*. Bloomington: Indiana University Press.

Thomas, F.C. (2004). *To the Mouths of the Ganges. An Ecological and Cultural Journey*. Norwalk: EastBridge.

Tinker, H. (1968). *Foundations of Local Self-Government in India, Pakistan, and Burma*. 2nd edn, New York: Praeger.

Tomalin, E. (2009). *Biodivinity and Biodiversity. The Limits to Religious Environmentalism*. Farnham: Ashgate.

Trevithick, A. (2006). *The Revival of Buddhist Pilgrimage at Bodh Gaya (1811–1949)*. Delhi etc.: Motilal Banarsidass.

Tucker, R. (2012). *A Forest History of India*, New Delhi: Sage.

Turner, D. (1981). Narborough Bone Mill. *Journal of the Norfolk Industrial Archaeology Society*, 3(1), online, www.norfolkmills.co.uk/Watermills/narborough-bone-mill.html (accessed 19 April 2014).

Varady, R.G. (2010). Land Use and Environmental Change in the Gangetic Plain: Nineteenth-Century Human Activity in the Banaras Region. In *Culture and Power in Banaras. Community, Performance and Environment 1800–1980*, ed. S.B. Freitag, 229–45. 2nd edn, New Delhi: Oxford University Press.

Venkat, A. (2011). *Environmental Law & Policy*, New Delhi: PHI Learning.

Vinten-Johansen, P. (2003). *Cholera, Chloroform, and the Science of Medicine: A Life of John Snow*. Oxford: Oxford University Press.

Waller, J. (2004). *The Discovery of the Germ*. Cambridge: Icon Books.

Watt, C.A. and M. Mann, eds (2011). *'Civilizing Missions' in Colonial and Postcolonial South Asia: From Improvement to Development*. London: Anthem.

Watts, S. (2001). From Rapid Change to Stasis: Official Responses to Cholera in British-Ruled India and Egypt: 1860 to *c*.1921. *Journal of World History*, 12(2): 321–74.

Wohl, A. (1983). *Endangered Lives. Public Health in Victorian Britain*. London etc.: J.M. Dent & Sons Ltd.

Worboys, M. (2000). *Spreading Germs. Disease Theories and Medical Practice in Britain, 1865–1900*. Cambridge: Cambridge University Press.

Worster, D. (1990). Transformations of the Earth. Toward an Agroecological Perspective in History. *Journal of American History*, 76: 1087–106.

Worster, D. (1988). Appendix: Doing Environmental History. In *The Ends of the Earth. Perspectives on Modern Environmental History*, ed. D. Worster, 289–307. Cambridge: Cambridge University Press.

Zühlke, L. (2013). *Verehrung und Verschmutzung des Ganges. Zusammenhang der ökologischen Probleme und der religiösen Bedeutung des heiligen Flusses*. Berlin: regiospectra Verlag.

Index

Page numbers in *italics* denote tables, those in **bold** denote figures.